Responsive Architecture

Responsive Architecture

Special Issue Editors

Dusan Katunsky
Jeffrey Huang

MDPI • Basel • Beijing • Wuhan • Barcelona • Belgrade

MDPI

Special Issue Editors

Dusan Katunsky
Technical University of Kosice
Slovakia

Jeffrey Huang
EPFL
Switzerland

Editorial Office
MDPI
St. Alban-Anlage 66
4052 Basel, Switzerland

This is a reprint of articles from the Special Issue published online in the open access journal *Buildings* (ISSN 2075-5309) in 2019 (available at: https://www.mdpi.com/journal/buildings/special_issues/ Responsive_Architecture).

For citation purposes, cite each article independently as indicated on the article page online and as indicated below:

LastName, A.A.; LastName, B.B.; LastName, C.C. Article Title. *Journal Name* **Year**, *Article Number*, Page Range.

ISBN 978-3-03921-698-7 (Pbk)
ISBN 978-3-03921-699-4 (PDF)

Contents

About the Special Issue Editors

Dusan Katunsky completed high school with a general education before graduating from the Faculty of Civil Engineering of the Technical University in Košice in the field of architectural engineering. He received his doctorate (PhD) from Slovak Technical University in Bratislava. He was Associate Professor at the Department of Building Structures of Slovak Technical University with a focus on building physics. In 2008, he was awarded the title of University Professor in the field of architectural engineering in Bratislava. He has worked as Director of the Institute of Architectural Engineering at the Technical University of Košice, where he is currently the Vice-Dean of Scientific and Research Activities. He is the author of several books and scripts for students. He actively acts as a reviewer for many magazines that are found in WOS and Scopus databases. He is also the author and co-author of many scientific articles. As the head of scientific committees, he was a part of many national and international scientific seminars and conferences. Currently, he is coordinating the bachelor, master, and doctoral study program within the field of the building structures and building physics at the Faculty of Civil Engineering, Technical University in Košice.

Jeffrey Huang is the Director of the Media x Design Laboratory and Full Professor at EPFL. He holds a DiplArch from ETH Zurich, and master's and doctoral degrees from Harvard University, where he was awarded the Gerald McCue medal for academic excellence. He started his academic career as a researcher at MIT's Sloan School of Management. In 1998, he returned to Harvard as Assistant Professor of Architecture and was promoted to Associate Professor in 2001. In 2006, he was named Full Professor at EPFL in Switzerland, where he holds joint professorships in Architecture and Computer Science, and heads the Media x Design Laboratory. During 2013–2016, while on leave from EPFL, he led the creation of a new school of architecture in Singapore (Bachelor's, Master's and PhD programs), as the Head of the Architecture and Sustainable Design Pillar at the Singapore University of Technology and Design (SUTD), a new university established in collaboration with MIT.

Preface to "Responsive Architecture"

Similarity or responsive architecture has been known for over 50 years. It arose in the middle of the last century and has many definitions. However, its basis is in the creation of buildings or building components that can adapt to external influences, external conditions, and the overall external environment. The building and its elements use physical states and force effects (light, heat, humidity, noise) in the form of sun, wind, and rain. These effects are used to change building elements or the building as a whole and to sensitively adapt in creating appropriate (self-sustaining, self-healing) sustainable architecture, buildings, and environments. The shape of the building will be able to adapt accordingly to the environmental conditions. External conditions also result in the adaptation of building elements to create a suitable indoor environment. The Special Issue "Responsive Architecture" in Buildings is dedicated to sensitive, creative, adaptable, and sustainable architecture. This question deals with case studies and demonstrations of specific (real) practical situations. It also deals with the design of adaptive building elements and structural justification of the design of adaptive architectural details of the building. Ideas that support the appropriateness of adaptive designs are also covered in this edition. The Guest Editors wish to thank the editorial team of Buildings for creating a space to address these very sensitive topics of architecture and engineering creation.

Dusan Katunsky, Jeffrey Huang
Special Issue Editors

![buildings logo] **buildings**

Article

The Influence of the Initial Condition in the Transient Thermal Field Simulation Inside a Wall

Marek Zozulák *, Marián Vertaľ and Dušan Katunský

Institute of Architectural Engineering, Faculty of Civil Engineering, Technical University of Kosice, 042 00 Kosice, Slovakia
* Correspondence: marek.zozulak@tuke.sk; Tel.: +421-155-602-4130

Received: 5 July 2019; Accepted: 26 July 2019; Published: 31 July 2019

Abstract: The envelope structures of buildings are exposed to heat-humidity conditions. The heat and humidity flow through these structures depends on the boundary conditions of the indoor and outdoor environments. This paper shows different initial conditions for the determination of temperature spread. The aim is to bring certain results of temperature calculated considering the initial conditions. When temperature changes, heat flow also rapidly changes. In certain specific cases, it is necessary to consider the initial conditions of temperature—for example; when transient energy simulations of real buildings are carried out. The reasons as to why it is necessary to consider the initial conditions, are shown in the examples of one-layer assemblies. The test walls exposed to the transient hygrothermal conditions are placed in the outdoor test cell. The cell has a stable temperature and relative humidity, and outdoor weather conditions change. The measured data on these walls and the calculated values of the temperatures in the wall structure, according to the different initial conditions, are compared. The average difference of the mean by the simulation and the measured values is significant. For a simulation time of about five days, the initial condition for calculating the temperature in the center of the masonry is necessary.

Keywords: boundary conditions for simulation; initial conditions; thermal conditions; heat flow; numerical analysis; experimental measurement; outdoor test cells

---

1. Introduction

We need to know the most exact situation of the initial boundary conditions for the calculation before we begin the computational simulation process. Analytical expressions, constants, or interpolation files are entered together with the properties of structural materials as input data, to determine the simulation criteria, initial conditions, and boundary settings [1,2]. Equilibrium simulation could calculate the distribution of internal air flow, distribution of air temperature and humidity in the room. These values are set as the initial transition analysis conditions [3]. A correct estimate of the initial conditions can have a significant impact on the predicted solutions as well as the overall actual condition of the building [4]. Initial temperature and air humidity must be specified when analyzing building constructions using simulation tools. The impact of the initial moisture content on the hygrothermal behavior of the building is investigated by many authors. The constant temperature of the built-in material in the building structure is generally considered to be the initial state of a rapidly changing factor. However, many simulation calculation tools use the initial temperature as a course of measured or predicted values. The aim is to show the importance of the preliminary calculation in the simulation process and the influence of the calculation on the temperature of the different types of building envelope materials. The preliminary calculation and its impact on the results of the simulation is compared to the measured temperature data, which is recorded in the experimental peripheral wall sections of the external test chamber. Numerical thermal field simulation is a calculation method used to obtain predictive temperature data that is validated using measured data in situ. In particular, the difference in measured and calculated temperature course values over time is quantified, depending on the determination of the initial temperature condition. The difference is influenced by the length of the analysed period.

As stated, the initial state of the transition temperature simulation typically represents the initial temperature distribution throughout the body profile at the beginning of the heat transfer process (t = 0). There are several ways to consider the initial value of the temperature in the structure according to its layout in the profile. The simplest assignment of simulation conditions is the use of a constant temperature across the analyzed component. A more accurate assignment of the initial conditions at t = 0 is the application of the measured temperature profile. Precise information on the temperature profile is rarely provided, so it is therefore appropriate to specify the initial state numerically. The temperature, thus determined, is calculated using the known standing ambient temperatures before time t = 0 or using a known transient temperature (initial pre-calculation). Mathematically, initial conditions are problems due to forward or inverse heat conduction. Problems with forward heat management aim to determine the temperature range of the medium, when the boundaries and initial conditions—the heat source / sink term (if any)—and the physical properties of the material are known [5]. On the other hand, the problem of inverse heat conduction concerns the estimation of the unknown initial temperature distribution, from the knowledge of the measured temperatures or the heat flux at time t > 0 [6]. In this study, the use of thermal initial conditions is demonstrated using numerical examples of simple structures—particularly experimental external test walls. Validation of the simulation is a valuable tool for checking the accuracy of numerical experiments. Outdoor test cells or climate chambers laboratories are suitable for such experiments, and our institution has already carried out similar research. The structure of state status monitoring under transient boundary conditions is the current trend in construction research [7]. An example of an experimental device for monitoring the physical properties of structures is a laboratory climate chamber, which has a "guarded" hot box [8]. Measurements can be made in existing buildings under conditions of use as described in [9]. In practice, it has proved difficult to measure all required inputs at the level required to obtain reliable performance estimates of components in real buildings [10]. Appropriate equipment—thanks to a controlled indoor environment—is an external test cell for the energy and hygrothermal assessment of structures and is for example implemented in Cottbus, Dübendorf, Glasgow, Limelette, Almería, Espoo, Delft, etc. Since 1993, outdoor test cells [11,12] have been employed in a wide variety of applications, from in-situ laboratory testing to complex building testing [12]. Cells are, among other things, useful for laboratory measurements of thermal comfort and emission efficiency, which are important factors in selecting heat emission systems in buildings with a low energy consumption, according to Kurnitski et al. [13]. Other institutes have extensive outdoor laboratories for research on buildings. They mainly focus on green architecture, green roofs and wooden lightweight building envelopes [14]. Another important point of focus in the field of building physics is the overheating of the building in summer. Outdoor monitoring devices as full-scale test cells could also be useful [15].

Accordingly, the external test cells were designed to: (i) form measurement data for hygrothermal reactions of the building envelope components in real climatic conditions in situ; and (ii) to form a validation data set for heat-air-moisture (HAM) simulations [10,16–18]. In the paper, the obtained test cell data are used to demonstrate their utility for initial and boundary applications when verifying dynamic simulations.

In order to work with the measured data, local climatic data should be collected at the location of the test cells in the exterior. The weather data used for the numerical calculation can be divided into: (i) typical weather data, (ii) weather data for project purposes, and (iii) weather data for individual investigations. In order to create a typical type of weather data, it is common practice to choose a typical year from a longer series of data according to the relevant criteria (reference climate year). Data for construction purposes are typically used to test the ability of a component to withstand the extreme conditions as they normally occur. During the investigation of the cause of the case, the weather data for each location and time period must be used. If measurements are made on a real building or in a building, calculations should be compared to verify the simulation model. In such cases, it is possible to use climate data from a nearby meteorological station [19,20].

The aim is to bring specific results of temperature in the fragments of structures according to different initial temperature considerations. Studies that would bring the certain numbers are missing. It is possible to look at the mathematical problem with the initial conditions, taking into account the certain values that we would bring. The results of temperature in the structures when using or not using the initial conditions are helpful to improve the whole simulation methodology.

2. Objectives and Methods

2.1. Measurement Experimental Test Setup

This experimental research—an experimental in situ measurement—was part of the completed research of experimental external test cells, which were an integral part of the Laboratory of the Faculty of Civil Engineering of the Technical University of Košice. The various opaque and transparent portions of the envelopes were monitored simultaneously in external and internal environments. The aim was to demonstrate the impact of transient natural boundary conditions on the internal structure and surface temperature, heat flow and moisture load [16,21].

Numerical analysis of the thermal field validated by experimental measurement is used to recognize the true heat behavior and moisture transfer in different kind of structures. Those analyses have been published in literature [22–27].

The measured parameters were: outdoor and indoor ambient air temperature and relative humidity; temperature and relative humidity inside the structure; surface temperature and heat flux. The room air temperature and relative air humidity were measured using a digital sensor, [28,29]. The weather station measured wind direction, wind speed, air pressure, air temperature and relative humidity, collisions, and also monitored global irradiation with Pyranometer, as shown in Figure 1.

Figure 1. Measuring equipment, instruments for measuring environmental values: (**a**) weather station; (**b**) pyranometer; and (**c**) temperature sensor and relative humidity of the air [28,29].

Sensor data collection is transmitted using a fully automated control panel which is connected to a USB cable and Ethernet cable on a computer, with an Internet connection. Data are retrieved and recorded in one-minute increments and stored on the hard disk of the computer.

The experiment was conducted on the (AAC) autoclaved aerated concrete and envelope structure from ceramic brick. The brickworks are thermally insulated using the thermal insulation contact system on the basis of graphite expanded polystyrene (EPS). The calculated thermal transmittance value of the opaque wall part is $U = 0.12$ (W / (m^{2}K)). The composition of the envelope wall can be seen in Figure 2 and Table 1. Results related to similar issues have been published in recent years in [30–33].

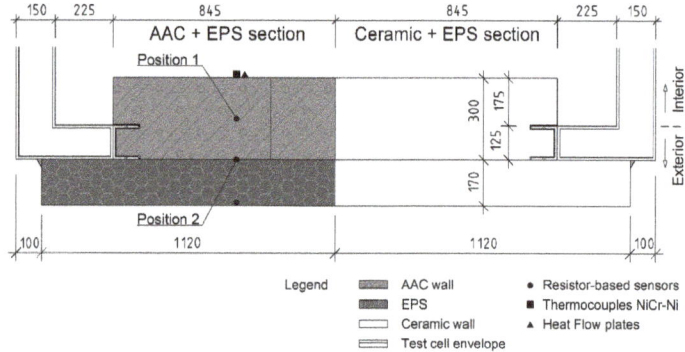

Figure 2. Configuration of the wall of the test cell and the position of the measuring sensors. Section view of the AAC section [28,29].

Table 1. Structure of the investigated perimeter wall (from the inside, the layers are in the direction of the heat flow) [28,29].

No.	Test-Wall Layer	d (m)	λ_D (W/(m·K))	c (J/(kg·K))	ρ (kg/m³)
1	AAC	0.300	0.104	900	350
2	Foam PUR	0.010	0.040	800	35
3	EPS polystyrene graphite	0.170	0.033	920	16
4	Adhesive mortar	0.002	0.850	900	1300
5	Primer basic paint	-	-	-	-
6	Silicone additive plaster	0.002	0.700	900	1700

2.2. Applicability of the Measured Data

The database of individual investigations consists of these measured parameters, from February 2012 to March 2015, and provides verification capabilities of various aspects: mathematical and statistical validation, non-stationary boundary conditions, material properties, and initial conditions for dynamic simulation. Any kind of simulation—such as that of energy, overheating, thermal fields, moisture load, or combined heat and moisture analysis—could be used.

All collected outdoor climatic data are used by authors for specific experimental purposes (on-site and numerical data), linked to the tested samples, and analyzed in external test cells for experiments in the locality of Košice, in the northern part of the city. Examples of external boundary conditions (BC)—which are mostly used in simulations—can be seen in Figure 3. Measured global solar radiation can be seen in Figure 4.

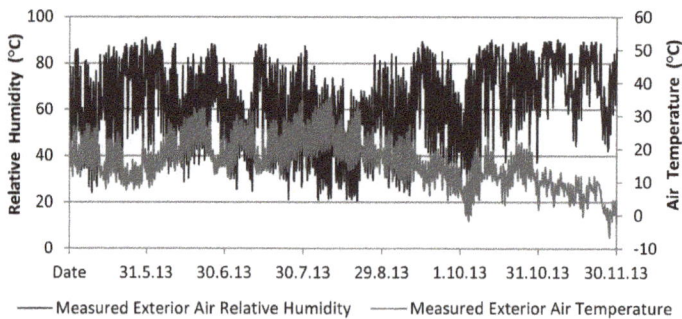

Figure 3. The measured outdoor air temperature in (°C) and the relative humidity of the air (%) for the locality of Košice. Displaying the selected period from May 2013 to December 2013 [28,29].

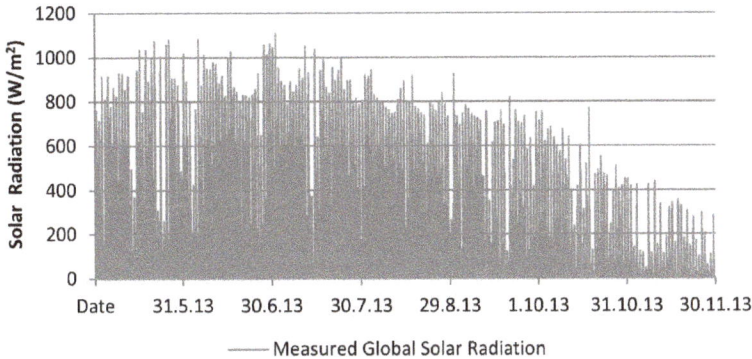

Figure 4. Global Solar Radiation (W/m²). Measured data for the locality of Košice. Displaying the selected period from May 2013 to December 2013 [28,29].

2.3. Transient Numerical Simulation

The heat transition over time is described by heat transmission (heat diffusion) in Equation (1). Thermal diffusivity is crucial for temperature balance in the structure in Equation (2). Equations (1) and (2) are described as follows:

$$div(\lambda gradT) + Z_q = c \cdot \rho \cdot \frac{\partial T}{\partial t}, \tag{1}$$

$$a = \frac{\lambda}{c \cdot \rho}. \tag{2}$$

The interim calculation is performed using the commercial software, Physibel, the BISTRA module. The method of energy balance is used to create a system of linear equations. The system is solved using the Crank-Nicolson finite difference method. This method meets the criteria of EN ISO 10211 Annex A for software computational methods [24,28,29].

2.4. Methodology for Determining Initial Temperature Conditions

Temperature equalization in simple walls (AAC and sandstone)—according to the different initial conditions used—is investigated. The structures of the wall samples embedded in the experimental test cell are shown in Figure 5. The measured, sinusoidal and constant ambient air temperatures over the selected time period as the boundary conditions are shown in Figure 6 and the measured values of internal and external surface temperatures for the analyzed time interval are shown in Figure 7. The initial temperature before the start of the calculation (before t = 0 hours (h)), can be considered:

- In the first case: in time $t = 0$, the constant temperature value 20 °C is set across the wall profile (denoted as IC 20 °C in Figures 8–17).
- In the second case: in time $t = 0$, the steady state temperature is calculated from the air temperatures: Exterior air is −13 °C, and interior air is 20 °C (BC −13 °C).
- In the third case: in time $t = 0$, the steady state temperature is calculated from the air temperatures: Exterior air is −3.42 °C, and interior air is 20 °C (BC −3.42 °C).
- In the fourth case: in time $t = 0$, the steady state temperature is calculated from the current ambient temperatures, without using start-up pre-calculation (BC without start-up).
- In the fifth case: start-up pre-calculation is used (calculation before $t = 0$ h, with a duration of one day), (BC start-up).

In the second case, we used an outside calculation (design) temperature of outside air for the city of Košice of −13 °C. The third case considers the outside temperature average during the coldest month of the year (January) for the location of the building, according to the standard [25].

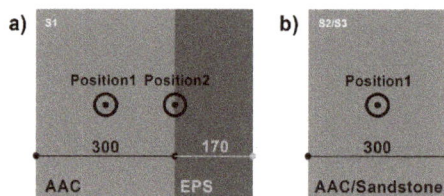

Figure 5. Sample structures, according to Table 2 (inside left). (**a**) An external test cell wall; and (**b**) simple AAC and sandstone walls. The placement position for the plotted data are: the calculated temperatures in Figures 8 and 10 in specimen S2/S3, in Figures 12 and 13 in S1 and the measured temperatures in Figures 15–17 in specimen S1.

2.5. Structure Specimens: Material Parameters

The physical properties of the material structures of the specimens are given in Table 2.

Table 2. Composition of the specimens (the layers are numbered from the inside in the direction of the heat flow) and physical properties of the material.

No.	Structure Specimens	Name of Layer	λ W/m·K	c J/kg·K	ρ kg/m^3
S1	Outdoor test cell wall (AAC + EPS)	AAC	0.106	900	350
		EPS	0.035	920	16
S2	Simple AAC wall	AAC	0.106	900	350
S3	Wall from sandstone	Sandstone	1.700	840	2600

2.6. Boundary Conditions

The exterior and interior temperature in the period from 15 to 20 February, 2015 are selected (in fact, from 14 February, 2015, for the purpose of the preliminary calculation). In Figure 6, the temperatures obtained from the weather data packet are selected and plotted. Simple sinusoidal functions for external and constant internal air temperatures used in simulation of sample walls are also plotted. In Figure 7, measured temperature values of the interior and exterior surfaces are plotted. These are used as the boundary conditions for validation with a full-scale experiment (Chapter 3.2).

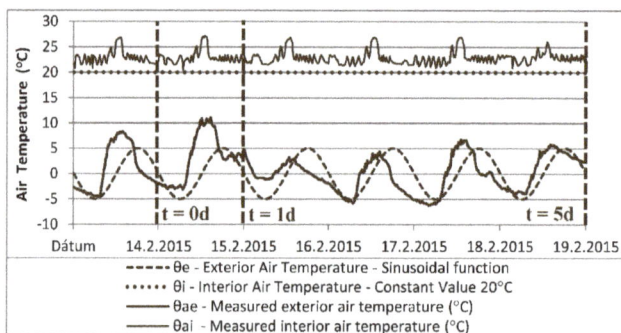

Figure 6. Measured, sinusoidal and constant ambient air temperature over the selected time period as the boundary condition used to determine the initial temperature conditions, as according to Section 2.4.

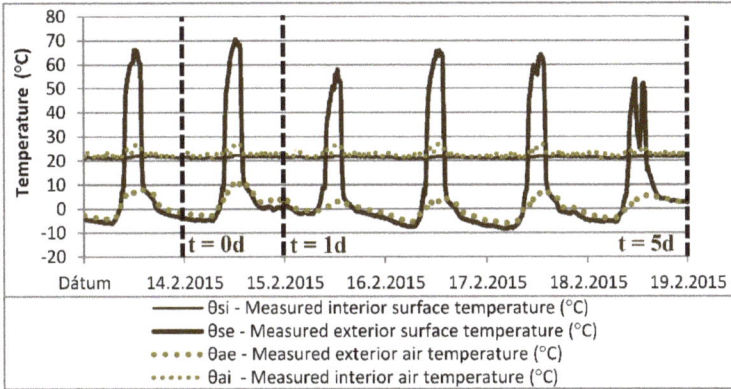

Figure 7. The measured temperature values of the interior and exterior surfaces for the analyzed interval (14 February 2015 to 19 February 2015).

3. Results and Discussion

3.1. Numerical Experiment Results

Figures 8 and 10 show the temperature balancing phenomenon due to a different thermal diffusivity (2) of the simple sandstone and AAC walls. The results of temperature represented by its courses, calculated in the test case using simple boundary conditions, consider various initial conditions. The boundary condition of exterior temperature for the calculation in this case is the sinusoidal function (θ_e, as shown in Figure 6). The boundary condition of interior temperature is a constant value of 20° over the presented time period (θ_i). Figure 9 shows sandstone; Figure 11 shows AAC; and Figure 14 shows AAC + EPS-simulated temperature profiles, after 0, 12 and 24 h. Temperature profiles are plotted to show their compliance with different IC considerations.

The temperature buffering time over the structure is extended when the thermal transmittance is decreased (application of external thermal insulation composite system, ETICS). When AAC with ETICS is used, temperature buffering takes five days or more (Figures 12 and 13). If the temperature course analysis in the structures has short time intervals, the correct initial condition is highly recommended and relevant.

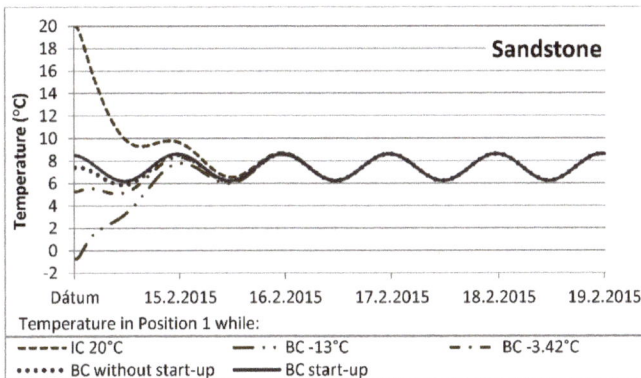

Figure 8. Plotted temperature courses of the simple sandstone test wall S3 in Position 1, according to Table 2 and Figure 5b during the chosen time period.

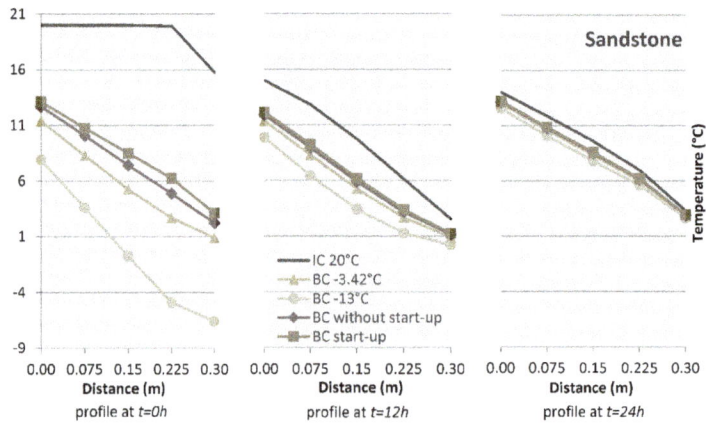

Figure 9. Temperature profile across the simple sandstone test wall at time $t = 0$ h, $t = 12$ h and $t = 24$ h for five cases under the initial condition. Abbreviations IC and BC refers to Section 2.4.

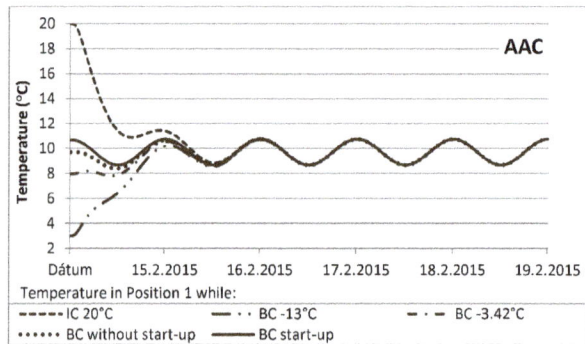

Figure 10. Plotted temperature courses of the simple AAC test wall S2 in Position 1, according to Table 2 and Figure 5b during the chosen time period.

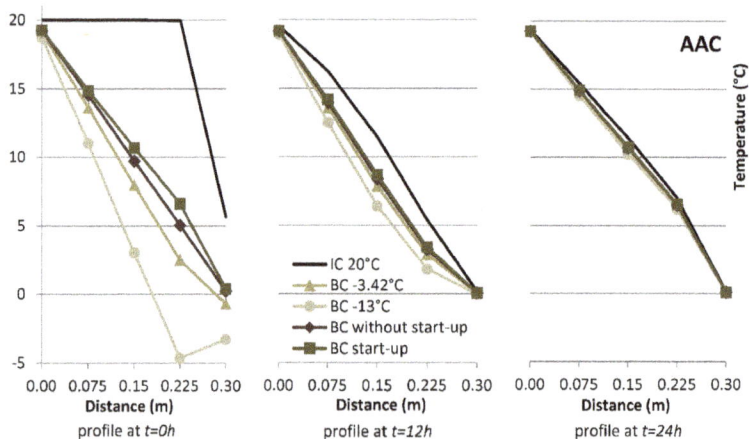

Figure 11. Temperature profile across the simple AAC test wall at time $t = 0$ h, $t = 12$ h and $t = 24$ h for five cases under the initial condition. Abbreviations IC and BC refers to Section 2.4.

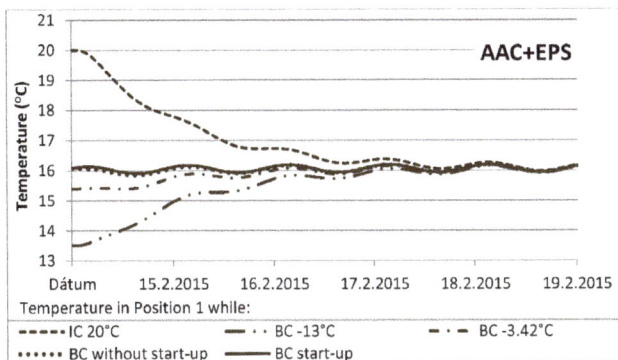

Figure 12. Plotted temperature courses of the outdoor test cell wall S1 in Position 1, according to Table 2 and Figure 5a during the chosen time period.

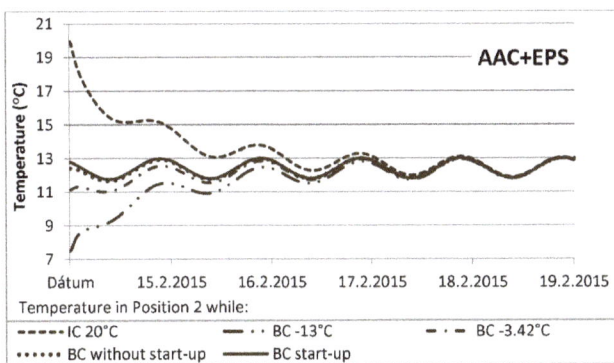

Figure 13. Plotted temperature courses of the outdoor test cell wall S1 in Position 2, according to Table 2 and Figure 5a during the chosen time period.

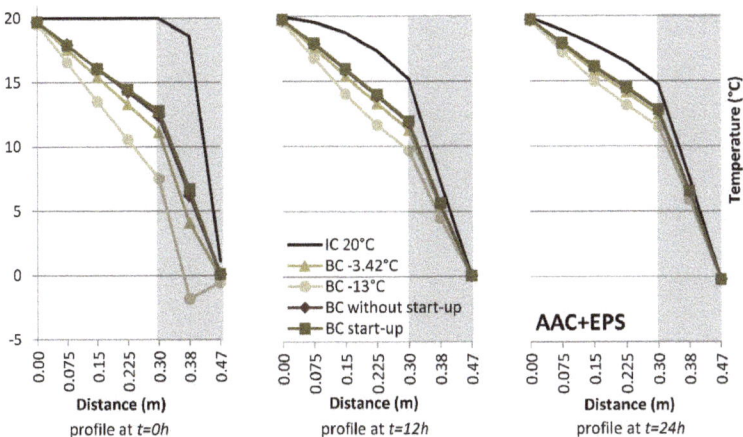

Figure 14. Temperature profile across the outdoor test cell wall (AAC + EPS) at time $t = 0$ h, $t = 12$ h and $t = 24$ h for five cases under the initial condition. Abbreviations IC and BC refers to Section 2.4.

9

The graphs of the temperature course (Figures 8, 10, 12 and 13) indicate that considering IC as the constant for the whole profile of the structure (IC 20 °C), or considering the calculated stationary profile using the standard outdoor temperature (exterior boundary conditions, BC −3.42 °C and BC −13 °C), causes uncertainty. During the simulation, the temperatures were gradually approaching the course calculated from the actual boundary conditions, with or without start-up pre-calculation (Chapter 3.2, fourth and fifth cases). In Chapter 3.2, in the validation of the numerical model of the measured data, we have considered only these initial conditions (fourth and fifth cases).

3.2. Model Validation with a Full-Scale On-Site Experiment

The impact of the initial condition is examined. The effect of solar irradiation is included by using measured temperature values of the interior and exterior surfaces as the boundary conditions for transient simulation (Figure 7). The influence of start-up pre-calculation on the speed of temperature balancing in the real case is considered. The data gained from the simulation are compared with the measured data. The data are investigated in the positions (1 and 2) of the mentioned test cell structure. Measured and simulated results from the simulation without pre-calculation and the simulation with one-day start-up pre-calculation are compared. Temperature courses in Figures 15 and 16—at the positions of the test cell wall—refer to the scheme in Figures 2 and 5.

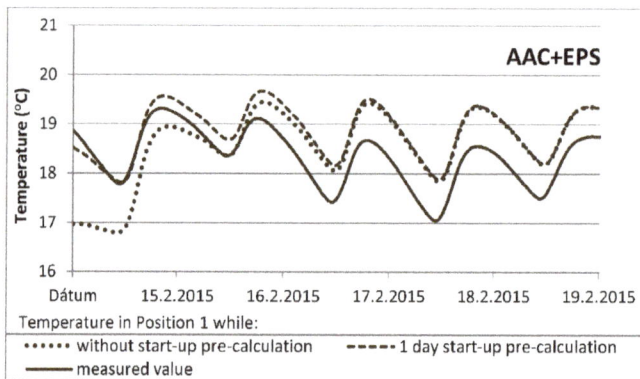

Figure 15. Plotted calculated and measured temperature courses in the outdoor test cell wall S1 in Position 1 during the chosen time period—with and without pre-calculation considerations.

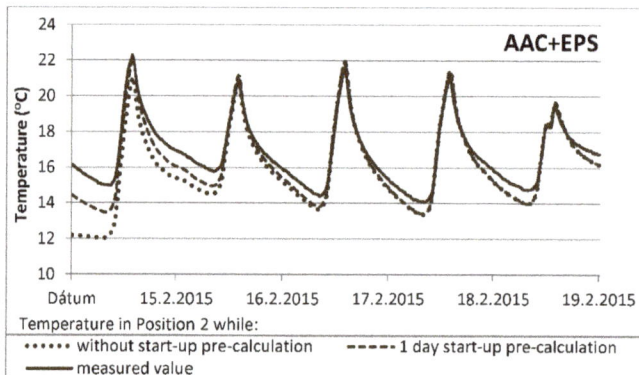

Figure 16. Plotted calculated and measured temperature courses in the outdoor test cell wall S1 in Position 2 during the chosen time period—with and without pre-calculation considerations.

The higher temperatures in position 2 (compared to position 1) are caused by the solar radiation effect (Figures 15 and 16).

Figure 17 presents a comparison of the temperature profiles across the testing wall, from the beginning of the simulation (0 h) for a time period of 24 and 48 h. A detailed analysis is provided in the discussion chapter.

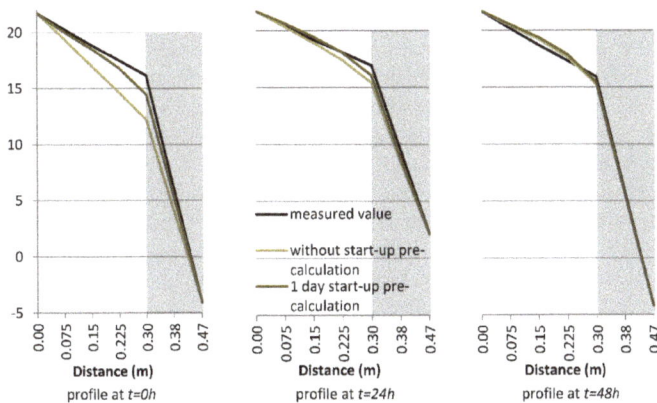

Figure 17. Temperature profile across the outdoor test cell wall S1 at $t = 0$ h, $t = 24$ h and $t = 48$—with and without pre-calculation considerations.

3.3. Discussion

Based on long-term analysis of thermal fields using non-stationary boundary conditions, the initial condition of temperature distribution in building structures can be partially omitted. The omission of the initial state condition can be considered to have been achieved when it has a short-term analysis of less than seven days. The rate of achieving the temperature balance and validity of the initial condition of the relevant calculation is given by the composition of the building structure.

To determine the initial condition correctly, knowledge of the temperature distribution of the cross section of the wall profile is required. However, this kind of information is rarely available. Accordingly, dissimilar simplifications need to be used when the physical analysis of buildings needs to be conducted. The article presents how the temperature behavior of the wall structure reacts to different initial conditions. Five cases—which are introduced in the text (Section 2.4)—are analyzed. The results of the calculation are displayed as time courses of temperatures at given points of the wall and temperature profiles across envelope structures.

The speed of temperature balancing in the building structure is, as expected, higher in terms of the one-layer structure than in the two-layer structure. The time required for temperature balancing does not exceed one and a half days for neither the material with a higher thermal diffusivity value (sandstone, Figure 8.), nor the material with a lower thermal diffusivity (AAC). After that, neither the method for determining the IC, nor the IC itself, had an influence on the distribution of temperatures in the analyzed building structure, as shown in Figure 10. The temperature balancing period is extended for the insulated structure, as shown in Figures 12 and 13.

Courses of temperature at selected points of the insulated envelope—as shown in schemes (Figure 5)—are compared in Figures 15 and 16. The temperature is calculated using start-up pre-calculation and then again without it, and the results are compared with the measured values. Temperature-data—calculated over one day of start-up pre-calculation—are a better match with the measured values within five days after the beginning of the evaluation process. At the beginning of the simulation, the difference between the measured and calculated temperatures for position 1 is only 0.2 K, and for position 2, the difference is 1.9 K. However, without start-up pre-calculation, the differences come to 1.9 K for position 1 and a much larger difference of 4.0 K for position 2. The average difference between the measured and calculated temperatures during the first 24 h of simulation without start-up pre-calculation is 0.9 K for position 1. For position 2, the difference is 2.2 K. Using start-up pre-calculation, the temperature differences are 0.1 K for position 1 and 1.0 K for position 2. Deviation between the measured and simulated temperature courses, especially in position 2, decreases during simulation. The average difference during a simulation interval, from day 2 to day 5, in measured and calculated temperatures without start-up pre-calculation for position 1 is 0.6 K, and for position 2, the difference is 0.5 K. Using start-up pre-calculation, it is 0.7 K for position 1 and to 0.3 K for position 2. This action on the course of the temperature (profile) across the envelope (Figure 17) is obvious; the calculated values are satisfactory due to the very accurate and exact application of start-up pre-calculation. The temperature equilibrium is observed after around five days of simulation. The influence of the IC consideration on the next results is negligible.

This analysis showed that the correct setting of the initial condition (characterizing the building structure at beginning of calculation) is important for the simulation of building structure behavior. By comparing measured with calculated courses, it is possible to answer the questions of whether individual factors have an influence on the entire accuracy of simulation.

Measured and calculated courses differ due to the neglect of certain factors that affect the heat transfer in the building envelope. One of these factors is the thermal field simulation mechanism used. The used factor does not include the complex HAM effects on the heat transfer process. The moisture content of the material aggravates its thermal conductivity coefficient and specific heat capacity. The simulation tool considers the comprehensive heat-air-moisture transport and allows for the inclusion of the regular initial condition, irrespective of whether the moisture, temperature or both may achieve the better compliance of the simulation and experimental onsite measurement [26,27].

4. Conclusions

The input data values imply the correctness of the numerical simulation. Consideration of the initial calculation conditions is more often used for the quantity, such as, e.g., the water content that persists in the material of the structure or any mass. Shorter lasting quantities—such as temperature— are often passed off as initial conditions in calculations. The article shows the possibilities of considering the initial temperature of structures as an initial calculation condition in transient simulations. The simulations result in various structures showing different initial temperatures reflected in temperature equalization. During several or just a few days, depending on the structure quality, distorted values are shown.

For the main result, in position 1, the temperature calculated using a pre-calculation and the temperature without a pre-calculation differ during the first three days of a five-day calculation by an average of 0.3 K. The average difference of the mean by the simulation and the measured values during the entire calculation period is up to 1.0 K and less. Therefore, for a simulation time of about five days, the initial condition for calculating the temperature in the center of the masonry is necessary. In position 2—that is, at the masonry interface and insulation—the temperature is calculated using the pre-calculation and also without the pre-calculation being differed for more than two days. The average difference is 0.4 K. Even at this point, the initial temperature condition of this wall composition is considered and is useful at this interval. The proof of the need for correct temperature IC is that when start-up pre-calculation is used, the simulation and measurement are compliant.

Further work should focus on the influence of the initial conditions on the thermal field for:

- Building structures with a complicated geometry, e.g., analysis of thermal bridges (2D, 3D)
- Building structure made of different building materials (window/wall connection), e.g., protruding building structures (cornice, pilaster, balcony, etc.)
- Building structure containing innovative thermal insulations with small building thicknesses and low values of thermal conductivity (reflective and vacuum insulation).

These experiments are feasible in the described outdoor test cells.

Buildings **2019**, *9*, 178

The article brings certain results of temperature in various fragments of structures according to different initial temperature consideration. Studies that would bring the certain numbers are missing. Due to the type of structure, different methods of initial condition considerations are needed. With the certain values that we bring, it is possible to look at the mathematical problem with the initial conditions taking into an account. The initial condition is a part of the input data in most of the HAM simulation programs. The point of view of building physics brings us general the conclusion that the temperature changes rapidly, and it is therefore not important to include it in calculations as an initial condition. In our climatic conditions (Slovakia, Middle-Europe) we insulate buildings to improve their energy efficiency, so most of the building envelopes consists of masonry and ETICs. Our study operates with initial conditions consideration in insulated structures. As we use HAM and BES or thermal field simulations, we need to be as accurate as possible. In our opinion, certain numbers of temperature in the structures when using or not using the initial conditions are helpful for improving our simulation methods, such as period and length of calculation, length of start-up pre-calculation—if it is used—and so on. Without answers regarding using or not using the initial conditions, it is impossible to gain reasonable, fair, accurate and right numbers for the results of temperature.

Author Contributions: Conceptualization, M.V. and M.Z.; methodology, M.V and M.Z.; software, M.Z.; validation, M.V. and M.Z.; formal analysis, M.V.; investigation, D.K., M.V. and M.Z.; resources, D.K., M.V. and M.Z.; data curation, M.V. and M.Z.; writing—original draft preparation, M.V. and M.Z.; writing—review and editing, D.K. and M.V.; visualization, M.V. and M.Z.; supervision, M.V. and D.K.; project administration, D.K.; funding acquisition, D.K.

Funding: This research was supported by the Slovak Scientific Grant Agency (VEGA) in collaboration with Slovak Ministry of Education, Science, Research and Sports and Slovak Academy of Sciences (SAS), under Grant number 1/0674/18.

Conflicts of Interest: The authors declare no conflict of interest.

Nomenclatures

AAC	autoclaved aerated concrete
EPS	expanded polystyrene
HAM	heat-air-moisture
ρ	bulk density, (kg/m^3)
c	specific heat capacity J/(kg·K)
λ	thermal conductivity W/(m·K)
a	thermal diffusivity (m^2/s)
q	heat flux (W/m^2)
Zq	heat source (W/m^3)
T	thermodynamic temperature (K)

References

1. Li, Q.; Rao, J.; Fazio, P. Development of HAM tool for building envelope analysis. *Build. Environ.* **2009**, *44*, 1065–1673. [CrossRef]
2. Šikula, O.; Mohelníková, J.; Plášek, J. Thermal analysis of light pipes for insulated flat roofs. *Energy Build.* **2014**, *85*, 436–444. [CrossRef]
3. Huang, H.; Kato, S.; Hu, R.; Ishida, Y. Development of new indices to assess the contribution of moisture sources to indoor humidity and application to optimization design: Proposal of CRI (H) and a transient simulation for the prediction of indoor humidity. *Build. Environ.* **2011**, *46*, 1817–1826. [CrossRef]
4. Woloszyn, M.; Rode, C. Tools for Performance Simulation of Heat, Air and Moisture Conditions of Whole Buildings. *Build. Simul.* **2008**, *1*, 5–24. [CrossRef]
5. Chiwiacowsky, L.D.; de Campos Velho, H.F.; Preto, A.J.; Stephany, S. Identifying Initial Condition in Heat Conduction Transfer by a Genetic Algorithm: A Parallel Approach. In Proceedings of the 24th Iberian Latin-american Congress on Computational Methods in Engineering (CILAMCE-2003), Ouro Preto (MG), Brazil, 29–31 October 2003; pp. 1–15.

6. Min, T.; Geng, B.; Ren, J. Inverse estimation of the initial condition for the heat equation. *Int. J. Pure Appl. Math.* **2013**, *82*, 581–593. [CrossRef]

7. Janssens, A.; Roels, S.; Vandaele, L. (Eds.) *Full Scale Test Facilities for Evaluation of Energy and Hygrothermal Performances*; International Workshop: Brussels, Belgium, 2011.

8. Martin, K.; Campos-Celador, A.; Escudero, C.; Gómez, I.; Sala, J.M. Analysis of a thermal bridge in a guarded hot box testing facility. *Energy Build.* **2012**, *50*, 139–149. [CrossRef]

9. Chalfoun, N.V. Using energy simulation and real-time data monitoring to investigate thermal performance of exterior cavity walls. In Proceedings of the 12th Conference of International Building Performance Simulation Association, (BS2011), Sydney, Australia, 14–16 November 2011; pp. 1211–1216.

10. Strachan, P.A.; Baker, P.H. Outdoor testing, analysis and modelling of building components. *Build. Environ.* **2008**, *43*, 127–128. [CrossRef]

11. Baker, P.H.; van Dijk, H.A.L. PASLINK and dynamic outdoor testing of building components. *Build. Environ.* **2008**, *43*, 143–151. [CrossRef]

12. Strachan, P.A.; Vandaele, L. Case studies of outdoor testing and analysis of building components. *Build. Environ.* **2008**, *43*, 129–142. [CrossRef]

13. Maivela, M.; Ferrantellia, A.; Kurnitski, J. Experimental determination of radiator, underfloor and air heating emission losses due to stratification and operative temperature variations. *Energy Build.* **2018**, *166*, 220–228. [CrossRef]

14. Ďurica, P.; Juráš, P.; Gašpierik, V.; Rybárik, J. Long-term Monitoring of Thermo-technical Properties of Lightweight Structures of External Walls Being Exposed to the Real Conditions. *Procedia Eng.* **2015**, *111*, 176–182. [CrossRef]

15. Kachkouch, S.; Benhamou, B.; Limam, K. Experimental Study of Roof's Passive Cooling Techniques for Energy Efficient Buildings in Marrakech climate. In Proceedings of the 2017 International Renewable & Sustainable Energy Conference (IRESC' 17), Tangier, Morocco, 4–7 December 2017.

16. Desta, T.Z.; Langmans, J.; Roels, S. Experimental data set for validation of HAM transport models of building envelopes. *Build. Environ.* **2011**, *46*, 1038–1046. [CrossRef]

17. Jiménez, M.J.; Madsen, H. Models for describing the thermal characteristics of building components. *Build. Environ.* **2008**, *43*, 152–162. [CrossRef]

18. Goia, F.; Schlemminger, C.; Gustavsen, A. The ZEB Test Cell Laboratory. A facility for characterization of building envelope systems under real outdoor conditions. *Energy Procedia* **2017**, *132*, 531–536. [CrossRef]

19. WUFI Creating Weather Files. Fraunhofer Institute for Building Physics. Last Update 1 November 2018. Available online: http://www.wufi.de/frame_en_wetterdaten.html (accessed on 16 June 2011).

20. Fokaides, P.A.; Kylili, A.; Kyriakides, I. Boundary Conditions Accuracy Effect on the Numerical Simulations of the Thermal Performance of Building Elements. *Energies* **2018**, *11*, 1520. [CrossRef]

21. Antonyova, A.; Korjenic, A.; Antony, P.; Korjenic, S.; Pavlušová, E.; Pavluš, M.; Bednar, T. Hygrothermal properties of building envelopes: Reliability of the effectiveness of energy saving. *Energy Build.* **2013**, *57*, 187–192. [CrossRef]

22. Katunský, D.; Zozulák, M.; Kondáš, K.; Šimiček, J. Numerical analysis and measurement results of a window sill. *Adv. Mater. Res.* **2014**, *899*, 147–150. [CrossRef]

23. Katunský, D.; Zozulák, M.; Vertaľ, M.; Šimiček, J. Experimentally Measured Boundary and Initial Conditions for Simulations. *Adv. Mater. Res.* **2014**, *1041*, 293–296. [CrossRef]

24. Physibel Software. *User Manual for Physibel*; Physibel Software: Maldegem, Belgium, 2017.

25. *Energy Performance of Buildings—Calculation of Energy Use for Space Heating and Cooling*; Standard STN EN ISO 13790/NA/Z1; International Organization for Standardization (ISO): Geneva, Switzerland, 2012.

26. Vertaľ, M.; Zozulák, M.; Vašková, A.; Korjenic, A. Hygrothermal initial condition for simulation process of green building structure. *Energy Build.* **2018**, *167*, 166–176. [CrossRef]

27. Vertaľ, M.; Vašková, A.; Korjenic, A.; Katunský, D. Fallstudie zum trocknungsverhalten von außenwandkonstruktionen aus porenbeton mit wärmedämmverbundsystem. *Bauphysik* **2016**, *38*, 378–388. [CrossRef]

28. Zozulák, M.; Vertaľ, M.; Katunský, D. Initial conditions determination for transient thermal field analysis. In *CER Comparative European Research 2014*; Sciemcee Publishing: London, UK, 2014; pp. 74–78, ISBN 978-0-9928772-24.

29. Zozulák, M.; Katunský, D. Numerical and experimental determination of in-structure temperature profiles. *Sel. Sci. Pap. J. Civ. Eng.* **2015**, *10*, 65–72. [CrossRef]
30. Shu, L.; Xiao, R.; Wen, Z.; Tao, Y.; Liu, P. Impact of Boundary Conditions on a Groundwater Heat Pump System Design in a Shallow and Thin Aquifer near the River. *Sustainability* **2017**, *9*, 797. [CrossRef]
31. Sarbu, I.; Iosif, A. Numerical Simulation of the Laminar Forced Convective Heat Transfer between Two Concentric Cylinders. *Computation* **2017**, *5*, 25. [CrossRef]
32. Yu, M.; Li, W.; Dong, B.; Chen, C.; Wang, X. Simulation for the Effects of Well Pressure and Initial Temperature on Methane Hydrate Dissociation. *Energies* **2018**, *11*, 1179. [CrossRef]
33. Wang, Z.; Wei, Y.; Qian, Y. Numerical Study on Entropy Generation in Thermal Convection with Differentially Discrete Heat Boundary Conditions. *Entropy* **2018**, *20*, 351. [CrossRef]

buildings

MDPI

Article

Quantifying the Generality and Adaptability of Building Layouts Using Weighted Graphs: The SAGA Method

Pieter Herthogs [1,2,*], Wim Debacker [3,4], Bige Tunçer [2,1], Yves De Weerdt [3] and Niels De Temmerman [5]

1   ETH Zürich, Singapore-ETH Centre (SEC), Future Cities Laboratory, 1 Create Way, Singapore 138602, Singapore
2   Informed Design Lab, Architecture and Sustainable Design, Singapore University of Technology and Design (SUTD), 8 Somapah Road, Singapore 487372, Singapore; bige_tuncer@sutd.edu.sg
3   Transition Platform, Flemish Institute for Technological Research (VITO), Boeretang 200, 2400 Mol, Belgium; wim.debacker@vito.be (W.D.); yves.deweerdt@vito.be (Y.D.W.)
4   Smart Energy and Built Environment, Flemish Institute for Technical Research (VITO), Boeretang 200, 2400 Mol, Belgium
5   Department of Architectural Engineering, Vrije Universiteit Brussel (VUB), Pleinlaan 2, 1050 Brussel, Belgium; niels.de.temmerman@vub.be
*   Correspondence: herthogs@arch.ethz.ch

Received: 28 February 2019; Accepted: 3 April 2019; Published: 20 April 2019

Abstract: This paper presents an assessment method that uses weighted graphs to quantify a building's capacity to support changes. The method is called Spatial Assessment of Generality and Adaptability (SAGA), and evaluates the generality (passive support for change) and adaptability (active support for change) of a building's spatial configuration. We put forward that the generality and adaptability of a floor plan can be expressed in terms of graph permeability, and introduce a set of five quantitative indicators. To illustrate the method, we evaluate six representative plan layouts, and discuss how their generality and adaptability scores relate to their spatial configuration. We are developing the SAGA method for two areas of application. First, SAGA's global graph indicators can be used to analyse and compare large sets of plan graphs, for example to map or plan adaptable capacity throughout a building or city. Second, the SAGA method can serve as a tool to inform design, allowing architects to improve the generality and adaptability of their plan layouts. While we conclude that the method has significant strengths and promising applications, the paper ends by discussing ways to make the assessment more robust and extend it beyond measuring spatial configuration.

Keywords: adaptability; generality; flexibility; evaluation tool; network analysis; Space Syntax; justified plan graphs; spatial analysis; architectural morphology; space plan

1. Introduction

Despite change being fundamental to cities and urban development, the built environment is not purposefully designed to support it. Buildings are often unable to accommodate changes over time [1] (p. 216); to keep up with changing standards, demands, or functional programs, they either need to be subjected to extensive renovations or demolished and replaced. This process is costly, resource intensive, and generates significant amounts of waste—construction and demolition represents an estimated 30 to 40% of global waste production [2] (p. 23). Moreover, the inability to support change can have impacts beyond the building level. Herthogs [3] hypothesised that the lack of 'adaptable capacity' of individual buildings (i.e. their passive ability to support change) introduces important inertias on the adaptive capacity of the urban ecosystem in its entirety (i.e. a city's active, emergent adaptivity), and put forward a methodology to study how adaptable capacity could be effectively distributed (i.e. planned) throughout the built environment.

Studying the benefits of 'building adaptability' requires a robust framework to quantify the various aspects of a building's capacity to adapt. However, although designing buildings for *adaptability* (or *flexibility*) has been 'a legitimate goal of architecture and planning' since the 1960s [4] (p. 51), it has arguably remained a niche field of study in architecture and building research. In a review paper discussing developments in building adaptability between 1990 and 2017, Heidrich et al. [5] (p. 296) discussed the lack of methods to score and evaluate adaptability, and emphasised the need to develop approaches and tools to inform design for adaptability. Geraedts et al. [6] (p. 1054) argued that existing adaptability assessment methods often only evaluate specific aspects of adaptability, such as technical or functional aspects, or focus on particular building types. Osman et al. [7] (p. 13) put forward that existing evaluations are not necessarily suited to assess buildings that were not designed to be adapted: obtaining results is often time-consuming and requires expert knowledge, but does not necessarily result in relevant data to inform conventional building design. In a review of existing assessment models, Rockow et al. [8] (p. 13) concluded that the modelling of building adaptation is 'in a nascent stage', and recommended the development of data-driven quantitative modelling approaches in future. They categorised the reviewed models ($n = 10$) according to their application: models either measure an existing building's potential for *adaptive reuse* ($n = 5$), a designed building's level of *design for adaptability* ($n = 4$), or the impact of adaptability on a building's life cycle ($n = 2$; one method combining design and life cycle assessment).

In this paper, we introduce a quantitative modelling approach called the SAGA method (Spatial Assessment of Generality and Adaptability). SAGA quantifies how well a building's spatial connectivity network can support change passively (i.e. *generality*) and actively (i.e. *adaptability*). Buildings with a high level of generality are designed in such a way that they can support changing needs and requirements without having to make physical alterations. Rather than being designed for a specific function, a general (or multi-purpose) building has characteristics—such as spatial layout, room sizes, or daylighting—that are suitable for general use. Adaptability is the result of (purposeful) decisions regarding design and detailing that make it easier to support changes in needs and requirements. This can be achieved using specific technological solutions, such as sliding walls, removable partitioning or reroutable service ducts, or by optimising the placement and hierarchy of conventional building components according to their technical or functional life spans (i.e. pace-layering [9]). Generality and adaptability represent the two dimensions of adaptability described by Heidrich et al. [5] (p. 288).

SAGA is related to the j-graph or Justified Plan Graph (JPG) method, an often-used graph analysis method for plan layouts. The JPG method is part of Space Syntax theory, a set of theories, methods and tools to analyse and understand the built environment in terms of spatial configuration and topology, primarily developed in the early 1980s in London [10].

The aim of the present paper is to introduce, explain, and illustrate the SAGA method's indicators for configurational generality and adaptability. The next section argues why graph permeability is a measure for the total number of functional uses a plan layout can support, which is the main premise of this work. Section 3 introduces three absolute indicators—*Generality* (G), *Adaptability* (A) and *Maximum Adaptability* (MA)—and two relative indicators—*Normalised Generality* (G_n) and *Normalised Adaptability* (A_n)—and discusses related literature. Section 4 discusses each step of a SAGA analysis in detail, explaining how to build plan graphs, calculate and normalise the indicators using *Permeability* (P). Section 5 illustrates the method by analysing six representative plan layouts, known to be general or adaptable, or designed to be. To conclude, we discuss the four main strengths of the SAGA method. Finally, we describe its potential to inform urban planning (by using its indicators to study and 'plan' adaptable capacity) and architectural design (by analysing and improving the generality and adaptability of an individual building layout).

The presented research has limitations. In terms of scope, SAGA emphasises plan analysis, which might not correctly represent more three-dimensional characteristics of a building layout. In addition, the method uses aggregate graph measures to quantify a multitude of potential uses, which likely results in a loss of sensitivity compared to conventional architectural or sociological plan analysis, and most certainly requires a shift in perspective. Moreover, the present work only considers configurational characteristics of a plan layout; although we are developing complementary indicators to evaluate different plan characteristics (such as surface areas), such characteristics are not yet represented. In terms of validation, the present paper is intended as a proof-of-concept of the proposed method. We argue that our main premise linking permeability to generality is theoretically sound, and we have applied the method to several representative cases (more than the six discussed in this paper). Nevertheless, a large-scale analysis covering a varied set of representative cases would be required to truly understand the practical potential and applicability of the indicators we introduced, and to fully prove the underlying premise. However, the level of proof is representative for the topic: in a review of existing adaptability assessment models [8], only one out of ten had been validated using a substantial dataset ($n > 100$); six out of ten were 'validated' using representative cases; three had not been validated.

We end the paper by suggesting how the calculation and evaluation could become more robust, and how the method could include a broader range of building characteristics related to generality and adaptability. We see SAGA as a spatial assessment framework with a range of assessment modules covering the spectrum of adaptable capacity, the assessment module introduced in this paper emphasising spatial configuration.

2. Premise: Permeable Plan Layouts are More General

To quantify the generality and adaptability of a building, SAGA starts at the level of the building layout or space plan. We argue that the properties of a building's space plan and its underlying building layout are at the core of a building's function, and strongly (but not solely) determine a building's capacity to support functional changes. This argument was illustrated by Schmidt III and Austin [11] (p. 57): their analysis of the interconnectivity of Brand's shearing layers of change [9] (i.e. Site, Structure, Skin, Services, Space Plan, and Stuff, extended to include Social, Space, and Surroundings) demonstrated that the Space Plan layer has both the highest number of links to other layers and the highest link strengths—see also: [5] (p. 289).

The SAGA method uses graphs to represent building layouts and their spatial configuration. In a *plan graph*, each space is represented by a vertex (or node), and each connection between two spaces is represented by an edge (see Figure 1). Both vertices and edges could contain additional information about the plan layout: vertices regarding the space they represent (e.g. its surface area, access to daylight, ...) and edges regarding the connection they represent (e.g. height differences, distances, ...). In the present paper, we adopt a purely configurational perspective: the plan graphs do not include any information regarding the size and location of spaces, or the physical distance between them.

A general plan layout can support changing uses without alterations. The use of a building—its *functional program*—can also be represented as a graph, with vertices representing spaces where particular functions take place, and (combinations of) edges representing the relational proximity between these different functions. From a strictly configurational perspective, ignoring requirements regarding the size or shape of spaces, one can therefore argue that the generality of a plan layout is directly related to the number of distinct graph representations of functional programs that fit within its plan graph. Hence, the potential to support different functional configurations is related to the overall *permeability* of a plan graph, as graphs with redundant edges can be traversed in multiple ways (they contain rings—i.e. have closed-loop paths). Steadman [12] (pp. 198–207), discussing 'adaptability and flexibility' in his book *Architectural Morphology*, provided a detailed worked example of this idea using a four-room plan, demonstrating that the total number of ways in which a functional program could be made to 'fit' inside a four-node plan graph increases with the graph's total number of edges *e*.

In other words, when a plan graph with v nodes is more permeable (or less linear), it can potentially shelter more unique v-room functional programs, and each of these programs can be organised within the plan graph in more unique ways—at least in a theoretical, configurational sense.

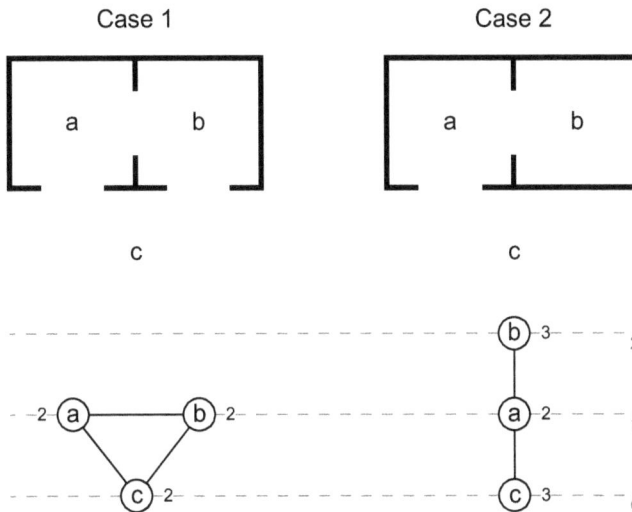

Figure 1. Plan graphs (below each plan layout) are graph representations of spatial configurations. The relations between the individual spaces (a, b, and c) in each case affect the overall spatial configuration of the plan layout. The Total Depth is listed next to each vertex.

While computing all configurational possibilities for a four-vertex graph is feasible, as the number of vertices increases the number of options quickly becomes computationally unwieldy. An alternative approach is to construct an indicator for permeability based on global graph characteristics, such as the *cyclomatic number* c ($c = e - v + 1$) [12] (pp. 189–191, 202–203), which counts the total number of closed faces in a graph (i.e. the number of rings that do not contain other vertices). The downside of a global graph measure is that it cannot measure the relative contribution of particular edges within the overall plan layout. Instead, we propose to quantify the permeability of a plan graph using *vertex depth*, which is the number of edges that need to be travelled to go from one particular vertex to another—usually a *root* vertex that has been singled out for analysis. In Figure 1, the plan graphs with root *c* are drawn under their respective plan layouts, with the distance from the root marked by grey horizontal lines; this type of drawing is a JPG. The *Total Depth* (TD) of a vertex—a measure used in JPG analysis—is the sum of its vertex depths to all other vertices in the graph. TD indicates how close or far a single space is to all other spaces, so the inverse of TD is a local measure of permeability. We then define *Aggregated Total Depth* (ATD) as the sum of the TD of every vertex, or the aggregated vertex depth from all vertices to all other vertices. In Figure 1, the fully permeable graph has an ATD of 6, while the linear graph has an ATD of 8. Rather than measuring a characteristic of the overall graph, ATD aggregates characteristics of individual vertices.

The main premise presented in this paper is that the generality of a plan layout increases with its permeability, and that this can be measured using (an inverse form of) ATD. Moreover, adaptability can be interpreted as the potential permeability achievable within a plan layout. The next section discusses these ideas at length. Afterwards, we explain how the ATD-based indicators can be normalised, how to construct plan graphs, and illustrate our premise using SAGA analyses of various plan layouts.

Initially, we developed the SAGA method as a reinterpretation and adaptation of the JPG method, one of the first analysis methods of Space Syntax theory. The JPG is featured extensively in its three seminal books [10,13,14]. For a detailed overview and critique of the method, consult [15]. The aim of the JPG is to study plan layouts in terms of configurations and relations rather than dimensions and geometry. It is used to study how spaces are related and connected, and whether there are correlations between patterns of spatial configuration and social behaviour or space use. The JPG applies and adapts graph theory to quantify these properties and identify important spaces within a building; an important form of analysis compares the integration (a measure of permeability derived from Total Depth) of individual graph nodes. Conversely, the SAGA method quantifies the permeability of an entire plan layout (i.e. the entire graph). Hence, characteristics related to generality and adaptability can be expressed in terms of a plan graph's ATD values. This implies a shift from local to aggregate vertex indicators, which explains the difference in interpretation and application between both methods: while the JPG method analyses the current use of one particular spatial configuration (i.e. one defined solution in the present time), the SAGA method analyses the potential future uses a spatial configuration could support (i.e. a solution space of hypothetical variants). Because SAGA uses aggregate measures instead of vertex measures, it hence foregoes social interpretation of a specific plan layout in favour of quantifying an unspecified number of potential uses. To avoid confusion between both methods, in the present paper we decided to express all indicators in terms of general graph theory and notation, and introduced a more appropriate normalisation method.

Nevertheless, the SAGA and JPG method are similar when it comes to graph representation, calculation and normalisation, and we often rely on JPG-related literature to illustrate particular points. Moreover, SAGA indicators can also be expressed in terms of JPG indicators (consult [16] for an overview). Two prior conference publications about the SAGA method [16,17] share similar narratives to the present article, but feature different calculation methods and cases; in both, the indicators were expressed in terms of JPG indicators (*integration* and *relative asymmetry*, respectively). Herthogs et al. [17] introduced the first results of the method: three indicators (G, A, and MA) applied to two variants of the same floor plan. The calculation method in [17] is incorrect: using average integration values results in a non-monotonic normalisation of the indicator (i.e. normalisation changes the ranking of cases). Herthogs et al. [16] didactically illustrated the main concept behind the SAGA method using two particular historic building types (one of which—the 'gentry house'—is also discussed in Section 5.1). The five indicators listed in [16] are based on JPG indicators; they produce different indicator scores, as the JPG method relies on a different *v*-node graph type for its upper bound normalisation.

3. Five Indicators for Configurational Generality and Adaptability

The main premise of the SAGA method is that the permeability of an existing plan layout is a measure of its *Generality* (G), with permeability expressed as an inverse of the Aggregated Total Depth (ATD) of the corresponding plan graph. As generality is a characteristic of an existing layout, the ATD is calculated using the *access graph* (*acc*) of the plan layout, i.e. a plan graph where edges represent physical connections between spaces (with a width of at least a standard doorway, or 0.8 m); this is illustrated in the first plan graph in Figure 2.

$$G \sim \frac{1}{ATD_{acc}} \tag{1}$$

As discussed in the previous section, a more permeable plan will have a higher probability to accommodate different functional organisations, as there are more possible ways in which its set of spaces can be connected and functionally arranged. Please note that this requires one to assume that if necessary, connections (doorways) can be closed when a floor plan is used in a different way.

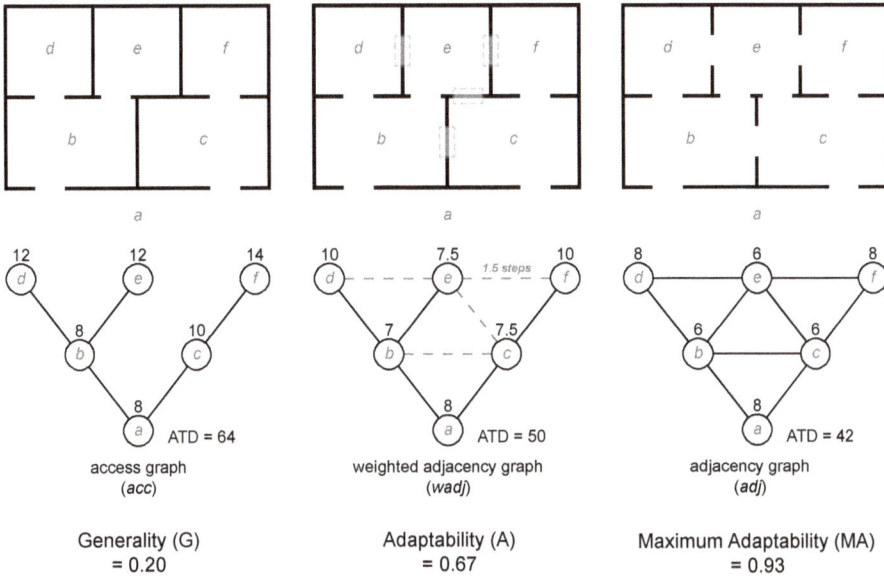

Figure 2. The SAGA method uses three related plan graphs to determine Generality, Adaptability, and Maximum Adaptability, which are proportional to the permeability of their respective plan graphs. Permeability can be expressed as the inverse of Aggregated Total Depth (ATD). For each of the graphs, the TD is listed above each node.

Several researchers have discussed the idea that a building's spatial configuration determines its capacity to support different uses. For example, Steadman [12] (pp. 198–207) explored the use of graph theory and measures related to permeability to quantify particular characteristics of the 'adaptability and flexibility' of plan layouts. Leupen [18] (p. 8) used plan graphs to explore and explain the generality of floor plans, arguing that the level of generality of a floor plan depends 'on the number of possible arrangements or combinations of activities it permits' [18] (p. 8). He stated that dwellings where every room can be accessed from a central, functionally neutral hub or via several different routes are easier to adapt to 'suit different living patterns' [18] (p. 1), and illustrated this by analysing five general layouts, all of which had star-like or ring-like spatial configurations, leading Leupen to conclude that these are essential characteristics for the generality of a floor plan [18] (pp. 8–9). Manum—e.g. refs. [19,20]—analysed several apartment types using the JPG method in order to understand their spatial configuration, surveyed the inhabitants of these apartments to establish types of usage patterns, and compared both to examine the 'potential usability' of housing types. One of the findings of his study was that apartment types with a less deep configuration are used by a wider range of household types [20] (p. 12).

The concept of adaptability can also be expressed in terms of plan graph permeability. In terms of spatial configuration, adaptability could be seen as the possibility to increase the permeability of a building, i.e. the ability to create new connections between rooms. In the access graph of a plan layout, spaces are either connected by a doorway or separated by a wall. If every possible connection between spaces would be made (by adding doorways wherever possible), the access graph would become the same as the *adjacency graph*, i.e. the plan graph that represents all rooms that are adjacent to each other (but not necessarily connected). Hence, the maximum permeability that can be achieved in a particular plan layout is proportional to the inverse of the ATD of its adjacency graph (*adj*). We call this indicator

Maximum Adaptability (MA); as with the Generality indicator, the MA indicator is inversely related to ATD.

$$MA \sim \frac{1}{ATD_{adj}} \tag{2}$$

Access graphs have binary edges: an edge either exists or not. Now imagine a connection that instead expresses a level of permeability: zero being impermeable, one being fully permeable, and values between zero and one expressing degrees of permeability. These degrees could represent the ease or 'probability' of creating new doorways. For example, it would be easier to connect two rooms separated by a stud wall than by a load-bearing brick wall.

We can represent this on a plan layout using a *weighted* adjacency graph. If we define the graph weight as the reciprocal of the probability, existing connections between rooms will have a weight (or vertex depth) of 1, while connections that pass through walls have a weight that is higher than 1—the inverse of a probability lower than 1 results in a graph weight that is higher than 1. In Figure 2, the second plan layout is identical to the first, except for four walls with an increased degree of permeability; instead of being impassable, it takes 1.5 steps to 'pass through' these walls. As a result, the TD of the individual nodes and the ATD is reduced. This implies a shift from vertex depth to shortest path calculations, from graph theory to network theory.

We put forward that the permeability of a plan's *weighted adjacency graph (wadj)* is a measure of the *Adaptability* (A) of the spatial configuration, with weights representing the difficulty to create a doorway between two unconnected rooms.

$$A \sim \frac{1}{ATD_{wadj}} \tag{3}$$

Existing connections have a permeability rating of one, while 'potential' connections have a permeability rating between zero and one. The resulting weights are the reciprocal of the permeability rating, resulting in connections that take more than one step to travel. Section 4.1.2 elaborates on the weighting of permeability.

We did not encounter prior studies using weighted plan graphs to directly measure adaptability. However, the basic idea is present in related work. For example, in an early study on the relation between building adaptability and the social use of space, Priemus [21] used weighted adjacency matrices to express and calculate the relationship between and clustering of different functions in a house (e.g. sitting, drinking coffee, doing laundry), with weights representing how closely related functions are. Based on surveys, Priemus also weighted user appreciation for walls with different degrees of permeability to separate these functions [21] (pp. 210–213). The canonical works of Space Syntax contain examples of spatial layouts with multiple possible graphs that represent a change in use: e.g. Hanson's analysis of the Rietveld-Schröder house [14] (chapter 7), or Hillier and Hanson's discussion of an Ashanti palace with special passageways for royals [10] (pp. 167–172). More recently, Behbahani et al. [22] used weights to represent characteristics of 'spatio-visual' relations between spaces, including the size of wall openings, to inform a discursive grammar method. Eloy [23,24] used JPGs to inform the development of a shape grammar for housing transformation and rehabilitation, with the aim to generate and evaluate alternative plan layouts for housing units that need to be adapted. The methodology included comparing the JPG of an original plan to those of refurbished variants, and labelling graph edges according to wall type (also in relation to their permeability). Several of her recommendations to improve a dwelling's 'flexibility' imply increasing or maintaining the permeability of internal walls.

G, A, and MA are absolute measures that can be used to compare different plan layouts. However, indicators G and A fail to take into account two important aspects related to the buildings these graphs represent. Firstly, the maximal permeability that can be achieved in a building with v spaces depends on the overall proportions of the building (layout). For example, a more rectangular building will likely have a higher ATD than a more square building, even if both have the same number of spaces and are fully permeable, as the proportions of a rectangular building are deeper and could restrict certain (shallower) configurations of spaces. Secondly, because the *wadj* is an extension of the *acc*, a high adaptability score could simply be the result of a building's very general spatial layout, and not the high permeability ratings of its internal walls.

Hence, we introduce two relative indicators that use MA as a benchmark for G and A—they can be used to determine how general or adaptable a floor plan is compared to the achievable minimum and maximum. Normalising G and A relative to MA results in the indicators *Normalised Generality* (G_n) and *Normalised Adaptability* (A_n), respectively. This concept is similar to a graph's *gamma index* [25], a global graph characteristic expressed as the number of edges in an access graph divided by the theoretical maximum edges in a planar graph (i.e. $e = 3v - 6$). However, unlike the depth-based SAGA indicators, the gamma index cannot measure the relative impact of the location of a vertex within the overall spatial configuration.

4. Step-by-Step Overview of a SAGA Analysis: Graph Representation, Calculation Methods and Conventions

This section describes the steps of a SAGA analysis: how to use its software tool, how to represent building layouts as graphs, how to rate the permeability of walls, and how to calculate and normalise the indicators. Table 1 provides an overview of the steps.

We developed a software tool combining *Rhinoceros 3D* [26], used to open a plan and draw convex maps and graphs, and its visual programming plugin *Grasshopper* [27], used to calculate all indicators and generate relevant colours, geometries, and vector images. Convex maps and graphs can be drawn in digital plans and models, or on top of images (e.g. photographs of archived building plans). Graph calculations have been implemented using Grasshopper plugin *Spiderweb* [28]. A floor plan can be scored within minutes. We refer to Figure 5 for an example of the output of the software tool.

Table 1. Step-by-step overview of the SAGA method.

Step 1: determine the plan's spaces		
Apply a modified convex mapping rule-set.		

Step 2: construct three permeability plan graphs		
access graph (*acc*)	weighted adjacency graph (*wadj*) with permeability scores	adjacency graph (*adj*)

Step 3: calculate the Aggregated Total Depth (ATD)		
ATD *acc* (vertex depth)	ATD *wadj* (shortest path)	ATD *adj* (vertex depth)

Step 4: normalise ATD to take into account graph size and relative permeability		
The **Permeability** (P) of a plan graph is:		
$$P = \left[ATD_{plangraph,v} - ATD_{linear,v}\right] / \left[ATD_{wheel,v} - ATD_{linear,v}\right]$$		
Generality (G) is the P of the *acc*	**Adaptability** (A) is the P of the *wadj*	**Maximum Adaptability** (MA) is the P of the *adj*

Step 5: normalize G and A relative to MA		
Normalised Generality (G_n): $$G_n = \frac{G}{MA}$$	**Normalised Adaptability** (A_n): $$A_n = \frac{A-G}{MA-G}$$	

4.1. Constructing Three Permeability Plan Graphs

4.1.1. Convex Mapping

To construct plan graphs consistently, the spaces in a building need to be formally defined. As with the JPG method, SAGA uses convex mapping to represent a plan layout as a plan graph. Traditionally, a plan graph is constructed using *convex spaces*: spaces in which every point can be seen from all other points. Convex spaces are representations of the visible space in a building [15] (p. 450), and do not necessarily correspond to functional rooms. Most notably, 'L-shaped' rooms are not convex and need to be split. Hillier and Hanson [10] (pp. 97–98) defined a rule-set stating that a non-convex space needs to be divided into the *fewest and fattest* convex spaces as possible, the former taking precedence over the latter, with fatness defined as the area-perimeter ratio of a convex space [10] (pp. 16–17).

The use of convex spaces can lead to a proliferation of the amount of spaces in the plan graph. This is especially the case in floor plans that have small setback spaces along walls due to columns, chimneys, alcoves, bay windows, etcetera. Ostwald [15] (p. 450) described how several authors do not use convex spaces to create JPGs, but use functional spaces instead. However, the SAGA method relies on convex spaces for the resolution of its indicators. Using functional spaces would result in a loss of information about the connections between spaces, as it becomes impossible to distinguish between the different walls of an L-shaped room. The density of (functionally useful) convex spaces determines the level of detail (i.e. number of edges) of the resulting graphs, and hence the resolution of the method.

To address particular issues of conventional convex mapping, we developed an alternative rule-set for SAGA. Firstly, we introduced a rule that allows the removal of convex spaces that are less than 0.8 m wide. Such spaces are not wide enough to accommodate a doorway, and can hence support neither movement nor occupation. When constructing convex maps, these spaces are simply ignored, and connections between two spaces that pass through them are replaced by direct connections. This results in far fewer convex spaces. Secondly, we proposed rules to make it easier to draw convex maps 'by hand'. We avoid the need to measure and compare perimeters iteratively by proposing a 'fewest and most connected' rule-set, which results in very similar convex maps. It does require several additional rules to guaranty consistency. In future, we plan to integrate methods to automate the construction of convex maps, e.g. refs. [29,30], removing what is arguably the most error prone human input.

4.1.2. Wall Permeability Ratings

The weights of potential connections in the adaptability graph are determined by permeability ratings, which express the probability of adding a doorway to a wall on a 0–1 scale. The permeability rating of a wall can depend on various criteria, such as whether it is structural or not, designed for disassembly, contains service ducts, etcetera. It is not the aim of this paper to introduce a comprehensive method to determine the permeability ratings of different wall types. The concept of an adaptability indicator can be adequately illustrated using a provisional set of values, as the concept of graph weights does not rely on the actual values of the weights.

We suggest provisional values based on the probability of creating doorways that are 0.8 metre wide; a conventional width that arguably represents a minimum comfort. We used a basic system of permeability ratings, shown in Table 2. There are three possible ratings: 0.3, 0.6 and 0.9, resulting in weights of 3.33, 1.66 and 1.11 respectively. There are also two modifiers that reduce the rating by 0.3, used if a wall has embedded plumbing, when it is load-bearing, or when it is a dividing wall. If a permeability rating becomes zero, the corresponding graph edge is removed. Walls with existing connections have a weight of 1 (fully permeable).

These provisional values only take into account constructional criteria. The list could be expanded to include, for example, functional and behavioural criteria, such as the current use of an adjacent space (e.g. whether a room is private or not), or criteria related to cost (e.g. whether a room houses specialised equipment or not). In future renditions of the method, we plan to implement a more elaborate framework to quantify wall permeabilities—e.g. refs. [31,32]. Both convex mapping and permeability weighting could be automated in a Building Information Modelling environment.

Table 2. Provisional values for wall permeability ratings (or probabilities) and modifiers.

Permeability Rating	
0.3	reinforced concrete wall
0.6	brick wall
0.9	removable wall (Designed for Disassembly)
Modifier	
−0.3	embedded plumbing
−0.3	load-bearing
−0.3	unit-dividing wall

4.2. Calculating and Normalising SAGA Indicators

4.2.1. Normalising Scores Relative to Graph Size and Type

The ATD of a plan graph is calculated by aggregating the vertex depths or shortest paths from all vertices to all other vertices. The ATD will naturally increase with the size of a plan graph, i.e. its number of vertices v. Hence, results need to be normalised for graph size to make them comparable. We normalise the ATD value of a particular v-size plan graph relative to the highest and lowest achievable ATD in a v-size graph. The JPG method has an analogous normalisation, transforming TD to *Relative Asymmetry*, which is a normalisation relative to its highest and lowest possible values—the end node of linear graph and the hub of a star graph, respectively [10].

The graph type with the highest ATD (i.e. the least permeable) is a linear graph. The ATD of a linear graph with v vertices is:

$$ATD_{linear,v} = 2v^2 - 6v + 4 \qquad (4)$$

The graph type with the lowest possible ATD (i.e. the most permeable) is not straightforward to determine. A complete graph, where all vertices are connected, would result in the lowest ATD, but is physically impossible for all but the smallest sized plan layouts: unlike vertices, rooms are not dimensionless. A *wheel graph*—i.e. a ring with a spoked central hub—is a highly permeable plan graph that is physically possible (e.g. interconnected rooms around a central room or corridor). The ATD of a wheel graph with v vertices is:

$$ATD_{wheel,v} = \frac{(v-1)^3}{3} + (v-1)^2 + \frac{2(v-1)}{3} \qquad (5)$$

We can then derive the *Permeability* (P) of a v-node plan graph by normalising its ATD relative to the minimum and maximum ATD achievable:

$$P = \frac{ATD_{plangraph,v} - ATD_{linear,v}}{ATD_{wheel,v} - ATD_{linear,v}} \qquad (6)$$

The Generality (G), Adaptability (A), and Maximum Adaptability (MA) of a plan graph are the Permeability of its access graph (*acc*), weighted adjacency graph (*wadj*), and adjacency graph (*adj*), respectively. For example, the Adaptability of the plan layout in Figure 2 is:

$$A = \frac{ATD_{wadj,v} - ATD_{linear,v}}{ATD_{wheel,v} - ATD_{linear,v}} = \frac{50 - 70}{40 - 70} = 0.67 \qquad (7)$$

These normalisations have a 0–1 range. Each of the three indicators express the percentage to which they achieve maximum permeability, i.e. how non-linear or wheel-like their plan graph is.

Please note that because these measures are normalisations, there is no need to invert ATD or average it per vertex (unlike in the JPG method).

Determining the most suitable graph type to represent the smallest achievable ATD warrants further study. While a wheel graph is highly permeable and physically possible, the likeliness of it occurring in the built environment will likely decrease as v increases. In buildings, the depth of the spatial configuration does not increase uniformly with the number of spaces in a building, but increases at a higher rate [10]. Larger buildings tend to be divided into units, sections, wings, or courts connected by corridors or passageways, resulting in multiple integrated clusters (i.e. units) linked by linear segregating spaces (i.e. the circulation between those units). In the JPG method, Relative Asymmetry values are further normalised to take this into account; in JPG-related literature, this normalisation process is subject to much debate and scrutiny, see e.g. ref. [33] (p. 8), ref. [34], ref. [15] (p. 453), which has led to several alternative approaches. Moreover, it is likely that particular building types tend towards particular graph configurations or overall dimensions. Perhaps such differences ought to be taken into account in SAGA's indicators, using different normalisation graph types for different building types. In future, we will likely devise a more general approach to normalise ATD; for now, comparisons between graphs of significantly different sizes will likely have some degree of normalisation bias.

4.2.2. Normalising Scores Relative to Maximum Adaptability

The final transformation, normalising G and A relative to MA, results in Normalised Generality (G_n) and Normalised Adaptability (A_n), respectively. They reflect how much Generality and Adaptability have been achieved in the plan layout relative to the maximum achievable Permeability score for that layout. The formulas are:

$$G_n = \frac{G - G_{min}}{MA - G_{min}} = \frac{G}{MA} \tag{8}$$

$$A_n = \frac{A - G}{MA - G} \tag{9}$$

The smallest possible Generality score is always zero (by definition, due to normalisation). G_n and A_n both have a 0–1 range.

5. Illustration: Six SAGA Analyses

This section illustrates the SAGA method by analysing plan layouts and discussing their indicator scores. Figures 3 and 5–8 show examples of SAGA analyses. In the figures, each convex space is represented by a node in its center, existing connections are drawn between nodes using black lines, and the connections of the adaptability graph are drawn in green, with their probabilities attached to the midpoints of the edge. Node colours are assigned based on a relative gradient from blue (highest Total Depth) to red (lowest Total Depth).

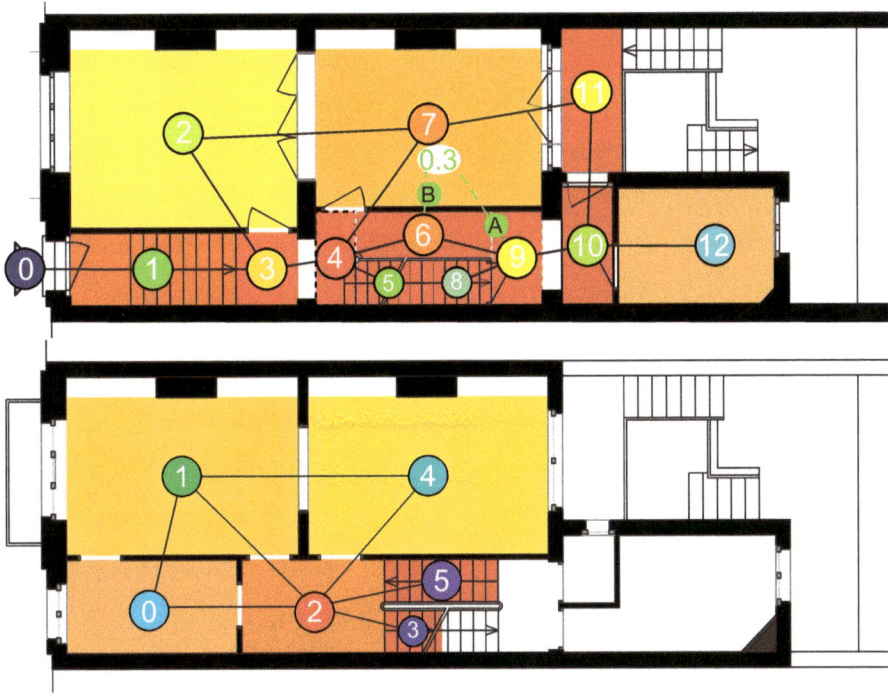

	v	*G*	*A*	*MA*	G_n	A_n
First floor	13	0.59	0.59	0.64	0.93	0.0
Second floor	6	0.80	0.80	0.80	1.0	n/a

Figure 3. SAGA analysis of two floors of a 19th century gentry house.

5.1. Belgian 'Gentry Houses'

The first two layouts are floors of a nineteenth-century 'gentry house' (French: *maison bourgeoise*; Dutch: *herenhuis*), a ubiquitous Belgian housing type. In Belgium, gentry houses are commonly considered to be highly general buildings. Figure 3 shows the SAGA analyses of the first two floors of a typical gentry house [35]—note that analyses are always done per floor level, by convention. Gentry houses reflect the lifestyle of the bourgeoisie in nineteenth-century Belgium, and had to support various household functions. While originally built for upper class families and their staff, today they have often been converted into apartment units, student housing, office spaces, or small commercial businesses, or still serve as a single-family dwelling. In Brussels, this building type represents about 25% of the residential housing stock—[35], authors' calculation.

A gentry house layout is organised perpendicularly to the street, featuring a circulation zone with staircases on one side, and two or three adjacent and connected spaces on the other. As a result, its spatial configuration typically has a high number of rings. While this is evident in the ground floor plan layout, its Generality score of 0.59 is not remarkably high, likely because this entry level floor has several linear connections to other levels. Worth noting is that this floor has an A_n of zero; there are only two walls that could be opened up, but their permeability rating is too low to influence ATD calculations. The spatial configuration of a gentry house is very specific, and has little opportunity for adaptation.

The SAGA indicator scores for the second floor illustrate the gentry house's defining feature in terms of Generality: all rooms of this standard plan are fully interconnected, resulting in a G, A, and MA value of 0.8. A typical house has two to four of such floors. There is no measurable Adaptability, as the access graph equals the adjacency graph (stairwells are impermeable by convention). For the same reason, the G_n is 1.0. This highly permeable plan layout, together with other characteristics—such as non-specialised rooms with ample surface areas, French doors, and good daylighting—helps to explain the highly versatile use of this building type.

5.2. Apartments Redesigned for Change

The next four layouts are apartment units from a proposal to refurbish a medium-rise social housing block, built in 1972, to improve its adaptability. Developed by Paduart [36], a key feature of this proposal is the use of removable and reusable walls, fully designed for disassembly. Figure 4 shows all units of Paduart's proposed 'redesign for change'; in this paper, we only discuss the SAGA scores of unit E (highest G), unit F (lowest G), unit A (lowest G_n), and a redesigned alternative version of unit A that improves its Generality.

Figure 4. Overview of the seven unit types in Paduart's refurbishment proposal. Image by Paduart [36], used with permission.

Apartment E (Figure 5) has the highest Generality of all the apartment types in the set (G = 0.69). All the spaces or functional clusters are connected to a neutral U-shaped hallway. Hence, spaces can be accessed independently, making the spatial configuration more permeable. Unit E is also the most Adaptable (A = 0.84). Because the kitchen, bathroom, toilet, and laundry have been directly connected to service risers, none of the nine potential new connections have a reduced permeability rating due to plumbing. The apartment layout achieves 80% of its Generality (G_n = 0.8), and 90% of its Adaptability (A_n = 0.9). Per comparison, Apartment F (Figure 6), which has a similar graph size, has the lowest Generality in the set (G = 0.51), but a relatively high Adaptability and Normalised Adaptability (0.76 and 0.84, respectively) due to the removable walls.

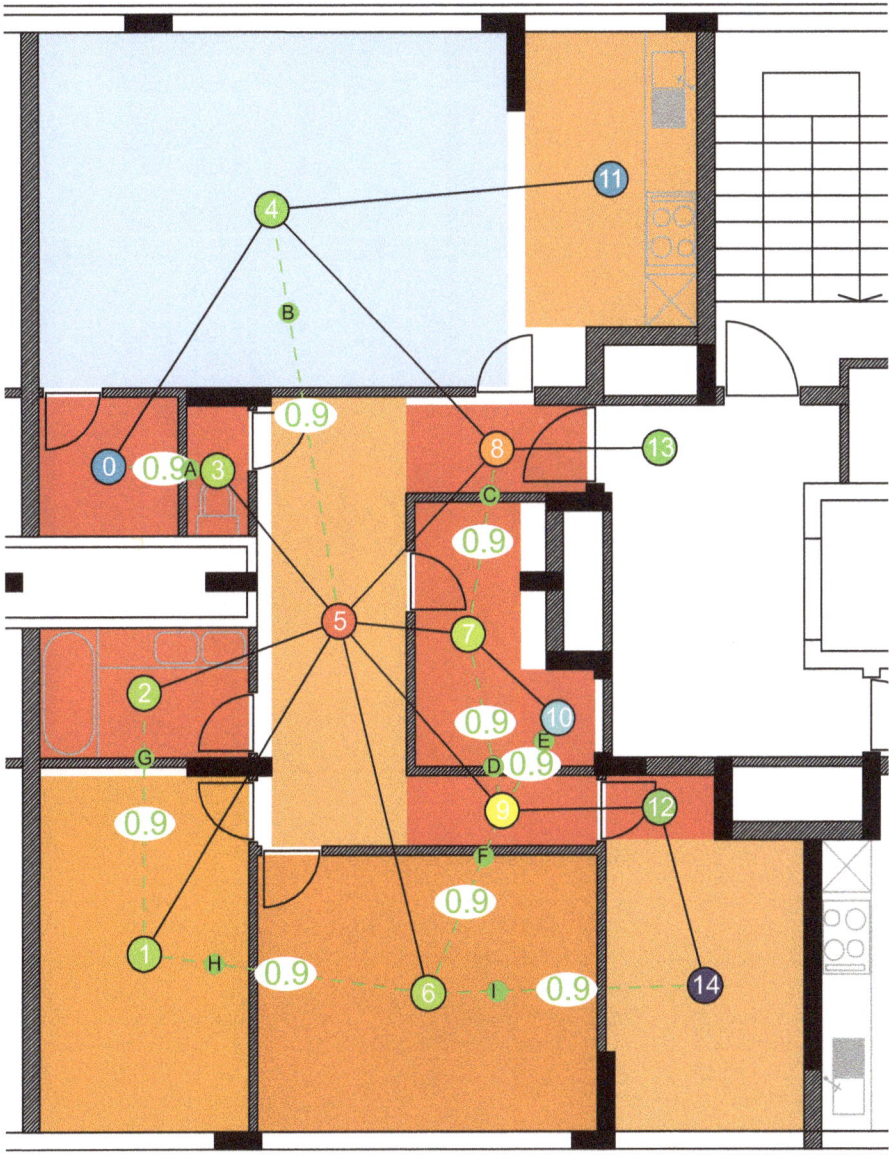

Apartment type E	v	G	A	MA	G_n	A_n
	15	0.69	0.84	0.85	0.80	0.90

Figure 5. SAGA analysis of apartment type E.

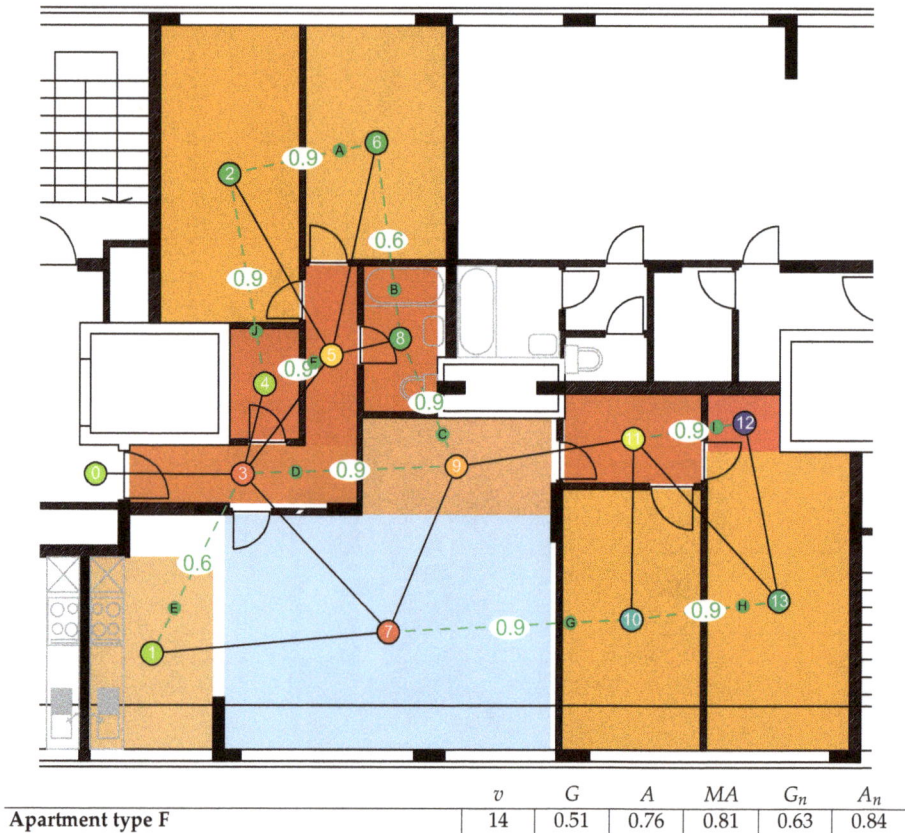

	v	G	A	MA	G_n	A_n
Apartment type F	14	0.51	0.76	0.81	0.63	0.84

Figure 6. SAGA analysis of apartment type F.

Figure 7 shows the SAGA analysis of apartment type A—a studio apartment. Its plan graph has the lowest Normalised Generality score (0.6) and the second lowest Generality (0.53) in the set. Almost all probabilities are 0.9 because all walls are removable (designed for disassembly), but the wall between the laundry and the toilet (edge D) contains plumbing (probability 0.6), and the wall between the hallway and kitchen is a demountable dividing wall (probability 0.6). Unit A has a low Generality score because the plan is mostly linear: the spaces in the back (toilet and bathroom, node 1 and 0) or the front (hallway and laundry, node 6, 5 and 4) are only connected through the large living space (node 3). However, Apartment A also has the highest Maximum Adaptability value in the set (0.89), indicating that Generality could be improved significantly.

Figure 8 shows an alternative layout for Apartment A, designed to improve its Generality. It extends the hallway through the laundry, linking up to the toilet and bathroom. The living space (blue) becomes accessible from two points, and the bathroom area can be accessed independently. The separate laundry is replaced by ceiling high storage along this extended hallway. A sliding door in the hallway gives users the option to shift between the less permeable layout of Apartment A, and the more permeable redesign. The rings introduced in this variant drastically increase G and G_n (to 0.72 and 0.84, respectively; both the respective highest values in the set). Also note the decline of Normalised Adaptability; there are less opportunities for adaptation in a general plan layout (as was the case in the gentry house). Nevertheless, the redesign of apartment type A demonstrates how a relatively small intervention (the removal of one key wall) can generate a significant increase in Generality.

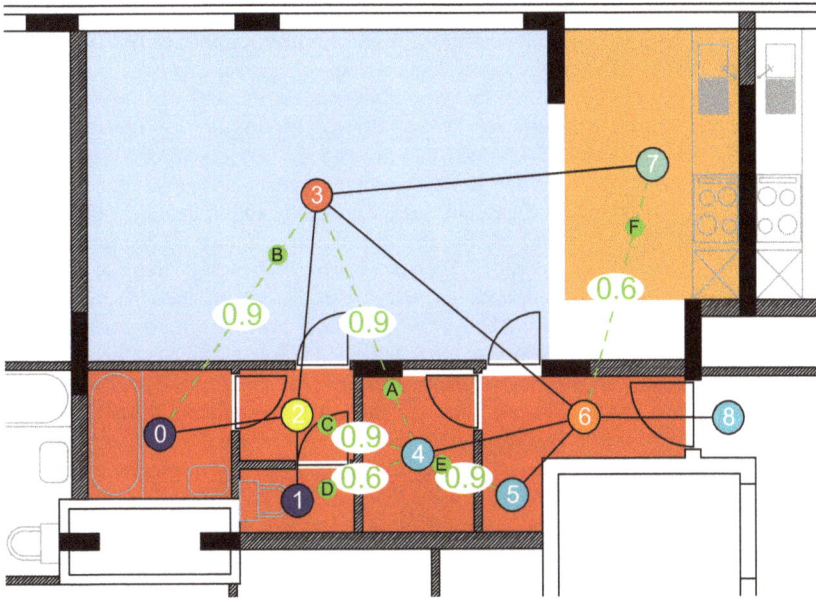

	v	G	A	MA	G_n	A_n
Apartment type A	9	0.53	0.80	0.89	0.60	0.76

Figure 7. SAGA analysis of apartment type A.

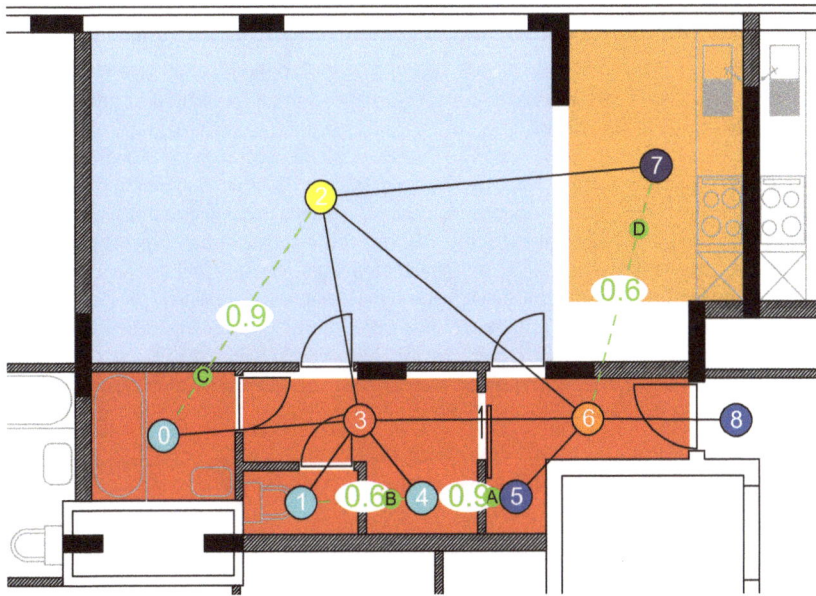

	v	G	A	MA	G_n	A_n
Apartment type A (redesigned)	9	0.72	0.80	0.86	0.84	0.58

Figure 8. SAGA analysis of apartment type A, redesigned for Generality by the authors.

6. Conclusions and Future Work

In this paper, we presented and illustrated SAGA, which is likely the first method to systematically quantify the generality and adaptability of building layouts using a combination of unweighted and weighted graphs. For that purpose, we introduced several indicators. SAGA has four important strengths. Firstly, it manages to quantify holistic design properties as a set of mathematically reproducible assessment indicators. Secondly, SAGA calculations are fast and do not require significant expertise or expert judgement, which lowers the threshold to evaluate a building's capacity to support change. Moreover, as the method is computational, it can be fully automated (e.g. within a Building Information Modelling environment). Automation could reduce human effort and error, but also enable data-driven assessment approaches. Thirdly, the method unifies the concepts of generality and adaptability in one mathematical framework, uncovering a direct relation between both, making them comparable. Fourthly, its indicators work both at the graph level, to evaluate the entire building layout, and the vertex level, to evaluate sub-parts of the layout.

The fourth strength translates to two main areas of application for SAGA. Firstly, its global indicators can be used to compare large sets of plan graphs (e.g. through sorting, ranking, clustering, ...), or to map or plan generality and adaptability throughout a large building or a neighbourhood. As explained in the introduction, the latter was the reason for SAGA's development [3]. Secondly, SAGA can be used for per-node analysis. Individual plans can be analysed to explain how local configurational features influence the global indicator scores. Hence, SAGA could also be a tool to inform design, allowing architects to improve the generality and adaptability of (parts of) their plan layouts. For example, SAGA can be used to calculate in which parts of a unit the use of removable walls would have the highest impact (i.e. highest increase in Generality).

Despite its novel features and strengths, there are several aspects that need to be improved or further developed. Firstly, the robustness of the method needs to be tested more thoroughly by applying SAGA to large numbers of floor plans, from diverse building types. We demonstrated the theoretical logic behind the main premise, that permeability is a measure for the potential number of uses of a building layout, and have illustrated our premise with a selection of cases. Although this level of validation is representative of the current state of adaptability assessment modelling [8], a systematic analysis of a large set of cases will help to validate the universal applicability of the method (by testing the basic hypotheses experimentally and empirically), to benchmark the indicators, and to establish a frame of reference to help interpret scores.

Secondly, the method should go beyond analysing spatial configurations, and needs to encompass more aspects of generality and adaptability. We see SAGA as a spatial assessment framework that will eventually feature a range of assessment modules; the results presented in this paper are the first step in its development. Additional indicator modules could address other aspects that determine generality and adaptability, such as floor surface area sizes (consult [16] for an initial version), wall permeability and related material reuse potential, the routing of technical services, wheelchair accessibility, or horizontal and vertical extendibility. All these examples could be measured using graph-based indicators.

Finally, SAGA's capacity to inform urban planning and architectural design practice ought to be explicitly studied and demonstrated.

Author Contributions: Conceptualisation, P.H.; methodology, P.H.; software, P.H.; formal analysis, P.H.; investigation, P.H.; writing—original draft preparation, P.H.; writing—review and editing, P.H., W.D., B.T., Y.D.W. and N.D.T.; visualisation, P.H.; supervision, N.D.T., Y.D.W., B.T. and W.D.; project administration, N.D.T. and Y.D.W.; funding acquisition, Y.D.W., N.D.T., W.D, P.H. and B.T.

Funding: Part of this research was funded by VITO (the Flemish Institute for Technological Research). Part of this research was conducted at the Future Cities Laboratory at the Singapore-ETH Centre, which was established collaboratively between ETH Zürich and Singapore's National Research Foundation (FI370074016) under its Campus for Research Excellence and Technological Enterprise program.

Acknowledgments: The authors would like to thank many reviewers for their constructive feedback. The first author would like to thank Anne Paduart for use of the building layouts she developed. The first author was a doctoral researcher at the Vrije Universiteit Brussel and VITO when the SAGA method was first developed.

Conflicts of Interest: The authors declare no conflict of interest. The funders had no role in the design of the study; in the collection, analyses, or interpretation of data; in the writing of the manuscript, or in the decision to publish the results.

Abbreviations

The following abbreviations are used in this manuscript:

SAGA	Spatial Assessment of Generality and Adaptability
ATD	Aggregated Total Depth
P	Permeability
JPG	Justified Plan Graph
TD	Total Depth
G	Generality
A	Adaptability
MA	Maximum Adaptability
G_n	Normalised Generality
A_n	Normalised Adaptability
acc	access graph
adj	adjacency graph
wadj	weighted adjacency graph
v	number of graph vertices
e	number of graph edges

References

1. Slaughter, E.S. Design strategies to increase building flexibility. *Build. Res. Inf.* **2001**, *29*, 208–217. [CrossRef]
2. Taipale, K. Challenges and ways forward in the urban sector. In *Sustainable Development in the 21st Century (SD21)*; Technical report; United Nations Department of Economic and Social Affairs (UNDESA): New York, NY, USA, 2012.
3. Herthogs, P. Enhancing the Adaptable Capacity of Urban Fragments: A Methodology to Integrate Design for Change in Sustainable Urban Projects. Ph.D. Thesis, Vrije Universiteit Brussel, Brussels, Belgium, 2016.
4. Hamdi, N. *Housing without Houses: Participation, Flexibility, Enablement*; Van Nostrand Reinhold: New York, NY, USA, 1991.
5. Heidrich, O.; Kamara, J.; Maltese, S.; Re Cecconi, F.; Dejaco, M.C. A critical review of the developments in building adaptability. *Int. J. Build. Pathol. Adapt.* **2017**, *35*, 284–303. [CrossRef]
6. Geraedts, R.P.; Remøy, H.T.; Hermans, M.H.; Van Rijn, E. Adaptive capacity of buildings: A determination method to promote flexible and sustainable construction. In Proceedings of the 25th World Congress of Architecture—Architecture Otherwhere, Durban, South Africa, 3–7 August 2014; pp. 1054–1068.
7. Osman, A.; Herthogs, P.; Davey, C. Are Open Building Principles Relevant in the South African Housing Sector? CSIR Investigations and Analysis of Housing Case Studies for Sustainable Building Transformation. In Proceedings of the International Conference on Management and Innovation for a Sustainable Built Environment, Delft, The Netherlands, 19–23 June 2011.
8. Rockow, Z.R.; Ross, B.; Black, A.K. Review of methods for evaluating adaptability of buildings. *Int. J. Build. Pathol. Adapt.* **2018**. [CrossRef]
9. Brand, S. *How Buildings Learn: What Happens after They're Built*; Penguin: London, UK, 1995.
10. Hillier, B.; Hanson, J. *The Social Logic of Space*; Cambridge University Press: Cambridge, UK, 1984.
11. Schmidt, R., III; Austin, S. *Adaptable Architecture: Theory and Practice*; Routledge: Abingdon, UK, 2016.
12. Steadman, J.P. *Architectural Morphology: An Introduction to the Geometry of Building Plans*; Pion Limited: London, UK, 1983.
13. Hillier, B. *Space Is the Machine: A Configurational Theory of Architecture*; Space Syntax: London, UK, 2007.
14. Hanson, J. *Decoding Homes and Houses*; Cambridge University Press: Cambridge, UK, 2003.

15. Ostwald, M.J. The Mathematics of Spatial Configuration: Revisiting, Revising and Critiquing Justified Plan Graph Theory. *Nexus Netw. J.* **2011**, *13*, 445–470. [CrossRef]
16. Herthogs, P.; Paduart, A.; Denis, F.; Tunçer, B. Evaluating the generality and adaptability of floor plans using the SAGA method: A didactic example based on the historical shophouse and gentry house types. In Proceedings of the UIA 2017 Seoul World Architects Congress, Seoul, Korea, 7–10 September 2017.
17. Herthogs, P.; De Temmerman, N.; De Weerdt, Y. Assessing the generality and adaptability of building layouts using justified plan graphs and weighted graphs: A proof of concept. In Proceedings of the Central Europe towards Sustainable Buildind, Prague, Czech Republic, 26–28 June 2013; Hájek, P., Tywoniak, J., Lupíšek, A., Růžička, J., Sojková, K., Eds.; Grada Publishing for Faculty of Civil Engineering, Czech Technical University in Prague: Prague, Czech Republic, 2013; pp. 992–998.
18. Leupen, B. Polyvalence, a concept for the sustainable dwelling. *Nord. J. Archit. Res.* **2006**, *19*, 24–31.
19. Manum, B. Apartment Layouts and Domestic Life: The Interior Space and Its Usability. Ph.D. Thesis, Oslo School of Architecture and Design, Oslo, Norway, 2006.
20. Manum, B. The Advantage of Generality: Dwellings' Potential for Housing Different Ways of Living. In Proceedings of the 7th International Space Syntax Symposium, Stockholm, Sweden, 8–11 June 2009; Koch, D., Marcus, L., Steen, J., Eds.; School of Architecture and the Built Environment KTH: Stockholm, Sweden, 2009; pp. 069.1–069.14.
21. Priemus, H. Wonen—Kreativiteit en Aanpassing: Onderzoek naar Voorwaarden voor Optimale Aanpassingsmogelijkheden in de Woningbouw. Ph.D. Thesis, TU Delft, Delft, The Netherlands, 1968.
22. Behbahani, P.A.; Gu, N.; Ostwald, M.J. Using Graphs to Capture Spatio-Visual Relations: Expanding the properties considered in Discursive Grammar. In Proceedings of the 47th International Conference of the Architectural Science Association, Hong Kong, China, 13–16 November 2013; pp. 187–196.
23. Eloy, S. A Transformation Grammar-Based Methodology for Housing Rehabilitation: Meeting Contemporary Functional and ICT Requirements. Ph.D. Thesis, Universidade Técnica de Lisboa, Lisbon, Portugal, 2012.
24. Eloy, S.; Duarte, J. A transformation-grammar-based methodology for the adaptation of existing housetypes: The case of the 'rabo-de-bacalhau'. *Environ.Plan. B Plan. Des.* **2015**, *42*, 775–800. [CrossRef]
25. Garrison, W.L.; Marble, D.F. *The Structure of Transportation Networks*; Technical Report TR 62-11; U.S. Army Transportation Research Command: Fort Detrick, MD, USA, 1962.
26. Rhinoceros. Rhinoceros (Version 5) Software. Seattle: Robert McNeel & Associates. 2018. Available online: https://www.rhino3d.com/ (accessed on 1 November 2018).
27. Grasshopper. Grasshopper (Version 1.0) Software. Seattle: Robert McNeel & Associates. 2018. Available online: https://www.grasshopper3d.com/ (accessed on 1 November 2018).
28. Schaffranek, R.; Vasku, M. Space Syntax for Generative Design: on the application of a new tool. In Proceedings of the Ninth International Space Syntax Symposium, Seoul, Korea, 31 October–3 November 2013; Kim, Y.O., Park, H.T., Seo, K.W., Eds.; Sejong University Press: Seoul, Korea, 2013; pp. 050:1–050:12.
29. Sileryte, R.; Cavic, L.; Beirao, J.N. Automated generation of versatile data model for analyzing urban architectural void. *Comput. Environ. Urban Syst.* **2017**, *66*, 130–144. [CrossRef]
30. Miranda Carranza, P. Convex maps, some basic concepts and a new method to generate them. In *Architectural Morphology: Investigative Modelling and Spatial Analysis*; KTH Royal Institute of Technology: Stockholm, Sweden, 2013; pp. 1–6.
31. Denis, F.; Vandervaeren, C.; De Temmerman, N. Using Network Analysis and BIM to Quantify the Impact of Design for Disassembly. *Buildings* **2018**, *8*, 113. [CrossRef]
32. Akinade, O.O.; Oyedele, L.O.; Bilal, M.; Ajayi, S.O.; Owolabi, H.A.; Alaka, H.A.; Bello, S.A. Waste minimisation through deconstruction: A BIM based Deconstructability Assessment Score (BIM-DAS). *Resour. Conserv. Recycl.* **2015**, *105*, 167–176. [CrossRef]
33. Livesey, G.E.; Donegan, A. Addressing normalisation in the pursuit of comparable integration. In Proceedings of the 4th International Space Syntax Symposium, London, UK, 17–19 June 2002; Hanson, J., Ed.; University College London: London, UK, 2003; pp. 64.1–64.10.
34. Park, H.T. Before integration: A critical review of integration measure in space syntax. In Proceedings of the 5th International Space Syntax Symposium, Delft, The Netherlands, 13–17 June 2005; van Nes, A., Ed.; TU Delft: Delft, The Netherlands, 2005; pp. 555–572.

35. B3RetroTool. Fiche Explicative: Maison Bourgeoise D'Avant-Guerres. Explanatory Sheet. 2016. Available online: https://www.brusselsretrofitxl.be/wp-content/uploads/2016/07/12_maison_bourgeoise_avant-guerres.pdf (accessed on 8 April 2019).
36. Paduart, A. Re-Design for Change. A 4 Dimensional Renovation Approach towards a Dynamic and Sustainable Building Stock. Ph.D. Thesis, Vrije Universiteit Brussel, Brussels, Belgium, 2012.

buildings

MDPI

Article

Thermal Responsive Performances of a Spanish Balcony-Based Vernacular Envelope

Isak Worre Foged

Department of Architecture, Design and Media Technology, Faculty of IT and Design, University of Aalborg, 9000 Aalborg, Denmark; iwfo@create.aau.dk

Received: 28 February 2019; Accepted: 4 April 2019; Published: 10 April 2019

Abstract: Many operable and complementary layers make up a vernacular adaptive envelope. With vertical operable translucent textile blinds, horizontal foldable glass doors with thin structural framing, wooden horizontal foldable frames with vertical rotational shutters, plants with dynamic densities, humidity concentrations, and opaque operable textile blinds forming the deep responsive façades of many Southern European buildings as part of the building envelope. This low-tech configuration utilizes behavioral human interaction with the building. On their own, these are singular mechanisms, but as coupled systems, they become highly advanced adaptive building systems used to balance temperature sensations. The research investigates such an adaptive envelope structure through identification of operable elements and their thermal and energy performances through computer simulation models. The designed research computational model includes assessment of heat reception and transfer, resultant operative temperatures, and adaptive comfort sensations. The aim of the research and the material presented in this paper is understanding the performance of native, local, low-tech systems as an opposing approach to contemporary high-tech, complex mechanical systems. The study finds that the operable elements and various compositions make a significant, yet less than anticipated, impact on adaptive thermal comfort temperatures.

Keywords: adaptive envelope; thermal performances; spanish balcony; simulation and computational design studies

1. Introduction

Architecture can be seen as a thermal interface, a membrane between the natural environment and humans. Depending on locality, the difference between thermal discomfort and comfort may be extreme or subtle, positioning the human in environments that are far from thermal balance, or hardly noticeable in respect to the physiology and homeostasis of humans [1]. At the same time, humans are dependent on the natural environment to survive, and hence, a form of synergy must be maintained. As humans are thermoregulatory animals, endotherms, acting dynamically in response to thermal and social stimuli, and most natural microclimates are in some form of continuous thermal fluctuation, architecture may, and perhaps should be, at the center point of acting, adapting and potentially balancing humans' correlations with microclimates together with humans. To an extent, with high certainty, IPCC [2] points to a future where microclimates will be more extreme in cold and warm conditions, local winds will increase in speed, and precipitation will become less predictable, requiring humans and buildings to become responsive to higher variation and unpredictability.

Vernacular architectures across continents are widely understood to include several strategies to act in response to local climatic conditions, which among others reduce thermal deviations from comfort temperatures, such as thermal storage, earth coupling, natural ventilation strategies, spatial programmatic layout, night cooling, vegetation, colouration, and more [3–6]. The strategies in focus here are operable elements for the regulation of solar radiation, which significantly contribute via the radiant temperature to the perceived operative temperature [7–9].

Adaptability, and particularly the adaptation of building envelopes, the principal regulatory membrane, could be central as a pathway to both increase local response capacities for humans and to decrease energy use in buildings by heat, ventilation, and air conditioning machinery, which currently largely is based on fossil fuel technologies. The relevance of this work to current research and building practice lies in the investigation to better understand how a low technological architectural envelope from vernacular architecture can become instrumental for future approaches to thermally-adaptive envelopes through thermal simulation studies.

Currently, in the context of thermally-responsive architecture, and specifically building envelopes, different approaches are taken in both academia and practice, with the objective to study material-intrinsic driven response and extrinsic-mechanical driven response processes [10,11]. In the built environment, as mapped by the International Energy Agency (IEA) Annex 44 reports [12] and [13] suggests that responsive building elements are segmented into everything from dynamic thermal storage to movable blinds. The survey by [10] and the IEA points to a near-singular approach of mechanical driven envelopes based on linear actuation, made as indirect control systems operated through electronic power supplies. In addition, the examples provided by these surveys point to one-layered responsive envelope elements. In academia, material studies are conducted to develop material-environmental driven responsive elements through hydrodynamics [14,15] and thermodynamics [16,17]. These studies aim to utilize energy that is available in the environment of a system to do work, also known as exergy systems [18–20], to actuate responsive behaviors. While the exergy-based responsive elements are effective energy systems, they are wholly relying on humidity and thermal changes in the environment, reducing the human behavioral adaptive capacities for response abilities. In contrast, if the human is considered both a decision-taking and actuation agent, she/he can act as an exergy-based adaptive system by engaging with the operable elements, which through interaction become responsive.

In contrast to a mechanical-based, extrinsic, indirect, and envelope single-layer approach commonly applied in industry, and a material-based, intrinsic, direct and envelope single-layer approach found in academic research, this study and approach to thermally-responsive systems in architecture is focused on extrinsic, direct actuation of multi-layer related elements by human actuation. This makes the human the center point of both sensing and actuation of the responsive system, rather than a passive agent, hence becoming an integrated adaptive feedback system. This approach to responsive building systems aligns with the adaptive comfort model principles of human engagement towards subjectively-experienced thermal sensations and the regulation of a local environment to meet these subjective sensations [21,22].

The background for this study is the analysis of what is here referenced as the Spanish balcony, not to be confused with the French Balcony, which is a mere fence and rail on the building facade, allowing normal size door opening of the envelope. The Spanish Balcony is used as case study for its design and instrumentality in allowing humans to create an adaptive and regulatory building envelope based on a series of simple elements, which combined provides for an advanced operable thermal architecture beyond the scale and operation of single architectural devices, as discussed among others by architectural theorist Leatherbarrow [23].

The background for future solutions is sought after and derived from the many operable and complementary layers that make-up a vernacular adaptive envelope. With vertical operable translucent textile blinds, horizontal foldable glass doors with thin structural framing, wooden horizontal foldable frames with vertical rotational shutters, and opaque operable textile blinds forming the deep responsive façade, many Southern European buildings have much more advanced thermal operable building envelopes than the single-layered approaches that are foregrounded today, as discussed above, and as seen in the Kiefer Façade by Ernst Giselbrecht and Partner (2007) and the Al Bahar Towers by Aedas Architects (2012). While the mentioned newly developed and applied systems are assumed to improve the thermal performances of the buildings as a whole, they remain in principle simple constructions, which are not operable by humans directly, thereby separating the building's climate actions from the occupier. In contrast, the low-tech configuration of the Spanish Balcony utilizes behavioral human interactions within the building, and in some cases dynamic plant properties and building components. On their own, they are singular mechanisms, but as coupled systems, they support the possibility for advanced adaptive building systems to balance temperature sensations.

The research investigates such an adaptive envelope structure through identification of operable elements and their thermal and energy performances through computer simulation models. The design research computational model includes an assessment of heat reception and transfer, and the resultant human-perceived adaptive thermal comfort.

The paper presents a simulation-based case study including a mapping of operable elements, and a simulation study including a parameterized computational model paired with a thermal-energy model and an evolutionary solver; the results of the computation-based testing and a discussion of the outlook follow this study. The study examines the thermal capacities of the Spanish Balcony to act as a thermal regulatory envelope by computationally searching for compositions of the operable elements to obtain thermal comfort as defined and calculated from the Adaptive Comfort Model [24,25]. The novelty of this work and the output includes both the thermal performances identified by the study and the proposition of adopting, adapting, and advancing multi-layered thermal adaptive envelopes for future envelope architectures.

2. Materials and Methods

The study methodology includes thermal modeling and evolutionary search procedures coupled through a geometric parametric model of the operable elements related to the Spanish Balcony.

2.1. Geometric and Material Model

The parametric model, developed in the software Rhinoceros Grasshopper by McNeel Inc, Figure 1, from inside to outside allows modification of white textile blinds mounted on inner wooden doors (A). The textiles can move along the door frame independent of the inner doors. The inner doors include standard single pane glass partitions (B). The doors are mounted on the inner side of the wall opening. Double-foldable doors/blinds are mounted on the outer side of the wall, where rotational movements are angle-symmetric so that the angle change between the first door partition and the second partition remains the same while operating in the model. For instance, when the first door partition (C) is rotated around its hinge 90 degrees, the second door partition (D) hinged to the first door partition is also rotated 90 degrees. In reality, the connected outer doors operate more or less independently of each other, restricted only in their rotation by the boundary of the balcony railing. Yet, the fixed angle relation is kept to reduce the amount of possible unique positions to reduce the vastness of the design search space. Within the outer doors are horizontal shutters positioned with the ability to rotate 180 degrees (E), allowing for the control of direct and indirect sunlight entering the internal space (F), which measures 3 m in height, 4 m in width, and 8 m in depth, with a 1.42 surface-to-volume ratio (136 m^2/96 m^3). All four doors operate independently. On the outside wall surface is a textile shading mounted marquise, which rotates around its hinge on the outer wall surface, allowing it to move freely, independent of the outer doors' positions (G). Internal partitions, including walls, the ceiling, and the floor are simulated as adiabatic. The outside facing wall is modeled as brick/cavity/brick with plaster (U-value 1.44 W/m^{2}K) [26] and glass in windows are modeled as a clear pane of 3 mm (U-value 5.88 W/m^{2}K). All operable element parameter values are normalized to create a search space with equally-sized dimensions. This is done to create a homogenous search environment and to reduce parameter variations while maintaining an estimated sufficiently large search space for combinations between the many different operable elements.

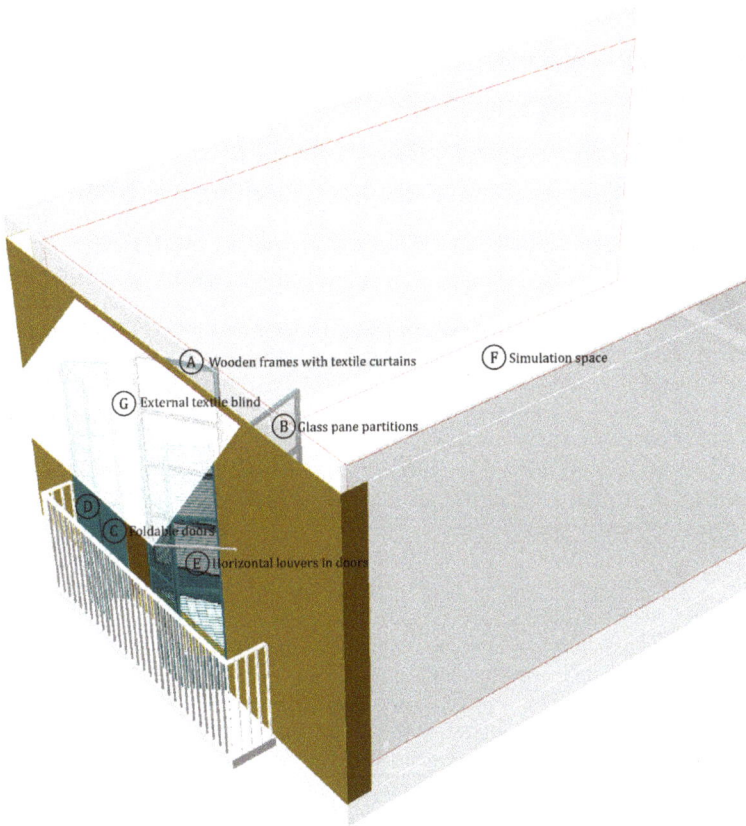

Figure 1. Perspective view of the parametric model including thermal and optimization simulations to search possible and suggested compositions of the multi-layered building envelope. The red line indicates the thermal zone simulated.

2.2. The Thermal Model: Understanding Adaptive Comfort Performances

The thermal simulation model is based on the Energy Plus simulation core and is integrated through Ladybug Tools as an add-on to Rhinoceros Grasshopper VPL. The thermal model is structured as a single-zone simulation with the above described thermal construction properties. The outer wall is modeled with three layers, including outer structural cladding, insulation (air cavity), and inner structural surfaces. The opening is facing South to an open field. All operable elements described above, including also the balcony geometries, act as shading elements in the simulation and thereby limit the heat transfer and direct sun radiation into the internal space. For each design variation of the operable elements, a thermal simulation is run full-year, hourly, focused on operative temperature. Based on the values from the full annual temperature analysis, an adaptive thermal comfort analysis is performed every 21st day of the month in the 12 months of the year. From this, a detailed thermal model is calculated, where the data from the selected days across the year are used as reference time periods from which the operable shading elements can be adjusted to obtain as high an adaptive thermal comfort as possible. The adaptive thermal comfort model is chosen over the heat balance model [27,28] as the operable elements seen with the Spanish balcony typically are modified by the occupant, actively engaging with the microclimatic conditions, thereby adaptively regulating subjectively the elements accessible to the occupant. All simulations use the climate data from the Energy Plus Weather Database for Barcelona, Spain, as a natural location for studying the Spanish balcony as a responsive system. Barcelona has mild winters and hot summers, classified as a warm-temperate subtropical climate, making the shading study particularly relevant, but also transferable to many similar climates globally.

2.3. Optimization Model: Searching Compositions

The optimization model is based on a Constrained Random Search (CRS) algorithm, namely the CRS2, implemented through the NLopt Library in the Rhinoceros Grasshopper VPL add-on Goat. This search algorithm is chosen as it performs a global search by the construction of random populations that then evolve, similar to evolutionary algorithms. While a brute force approach would be interesting to apply, finding thermal performances for all possible combinations, such survey and categorization lie outside the scope of this work. Instead, the optimization process is run for day 21 in each month. Each evolutionary simulation runs for 5 h, reaching approximately 50 iterations. At this level, in relation to the search space size, the search has not converged yet, resulted in proposing compositions of the operable elements where increased adaptive thermal comfort may be reached. The fitness function is aiming to reduce average deviations from thermal adaptive comfort across the 24 h of the day of the simulation. In this way, the method and model could also be used for searching compositions that at a given time of the year, down to a specific hour, would reach a particular subjective temperature of interest, which is different from the generalized comfort estimation otherwise applied in current practice.

2.4. Design Exploration

In this study, the design research experimentation does not lie within the exploration of elements and forms in their own right, but in how pre-established elements observed in the built environment can be composed to perform thermally for humans.

Hence, the study investigates possible designs of element behavior as a result of human engagement and adaption. This foreground design processes that attempt to include occupancy behavioral patterns as part of the design of elements, as these frame the space of possible configurations which eventually determine the quality of the design in respect to its thermal adaptive capacities.

From this, studies are conducted firstly to make and identify the operations possible with the elements available with respect to how these are related in their position and freedom to be configured on their own and as associated operable elements. As an example, in some observed cases of the Spanish Balcony by the author, the marquise is not able to be used before the outer doors are fully opened (or closed), which creates a dependency which decreases interoperation between the elements. In other cases, no marquise is applied, relying on the outer doors to provide shade. Such an approach would limit the ability for shading while maintaining a possibility for viewing between internal and external spaces. To allow a freer operation of elements and to maintain the possibility of shading while maintaining visual transparency through the building envelope, the marquise is included and is free to move independently of outer door positions. For the study to keep focus and to identify the thermal performances of the operable shading elements, other passive aspects, such as small plants on the balcony, thermal mass, and cross ventilation inside the building are not included in the simulations.

3. Results

From the simulation studies, data are extracted and visualized below. The data is structured in two studies and results. Firstly, how the adaptive comfort temperature is across the year, shown for the 21st day in each month for two conditions, an open and a closed envelope, Figures 2 and 3, is shown. Secondly, a series of compositions, Figure 4, are examined and visualized for the month, June, with the highest temperature deviation between open/closed conditions to understand the extent to which the operable elements can improve the thermal comfort as defined by the adaptive comfort model, Figure 5.

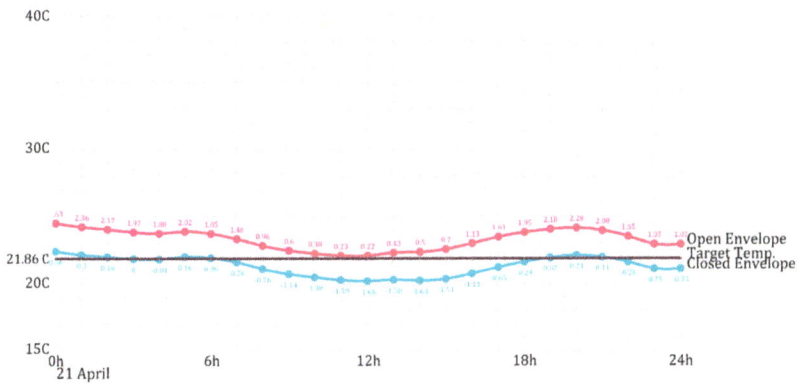

Figure 2. Temperature graph of 21 April showing thermal deviations from the calculated adaptive comfort temperature and generalized target temperature model in response to the operable element's composition. Graph y-axis includes temperatures from 15–40 degrees Celsius. Graph x-axis includes hours of the specific day, from 0–24 h.

Within the first 2 months of the year, the envelope in a fully open condition is close to the calculated target temperature (which varies according to the outdoor temperature). From months 3–10, an entirely closed envelope has less temperature deviation from the adaptive comfort target temperature. And, from months 11 and 12, the envelope in its fully open condition deviates the least. In the month of most deviation, July, temperatures reach 12 degrees above the target of 25 degrees Celsius, and the deviation between an open and closed envelope condition is simulated to only approximately 3.4 degrees Celsius. One aspect, which may cause part of this relatively little deviation is that thermal storage, a typical characteristic of older buildings with Spanish Balconies, due to stone/bricks being part of the core construction material is not accounted for. Another aspect is the thermal enclosure that is created by the closed envelope, limiting thermal exchange through convection.

When visualizing the data from the composition search, limited to 21 June, Figure 5, we can see that a series of compositions improve the gap between the comfort target temperature and the deviation temperature (purple color). What this suggests is that compositions can balance between open and closed envelope states to increase thermal comfort, yet, they do not manage on their own to reach comfort levels. For June, in Barcelona, the sun sits high, which also means that when the external marquise is just slightly extracted, the remaining operable elements will be placed in shade, limiting their contributions to regulate the direct solar radiation. In this study, this is particularly evident, as the envelope faces South, thereby receiving direct radiation in the middle of the day where the sun is at its highest position. The envelope design compositions related to the data in Figure 5 can be seen in Figure 4.

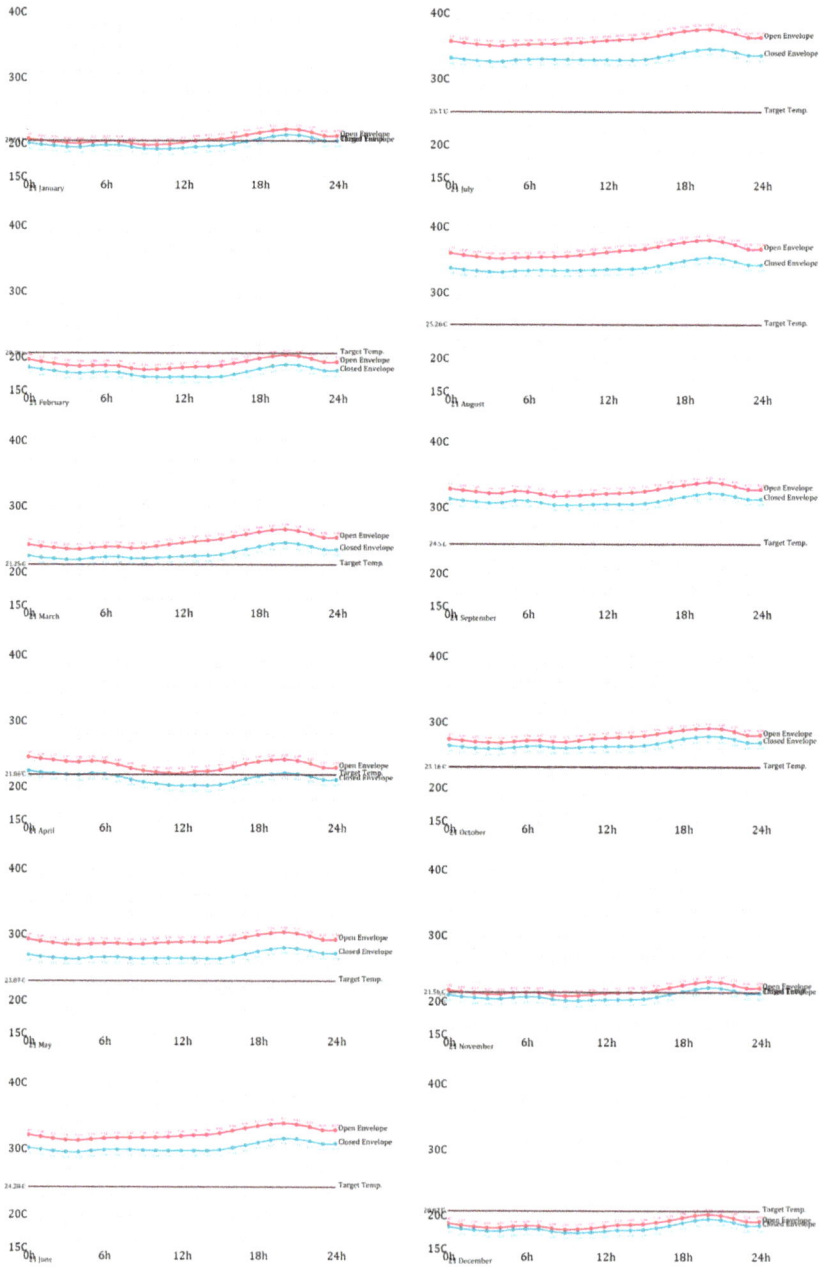

Figure 3. Temperature graphs of each of the 12 months showing thermal deviation from the calculated adaptive comfort temperature in response to the operable element's composition. Graph x-, y-axis values are kept the same as in Figure 3 to allow direct visual comparison between months.

Composition 6

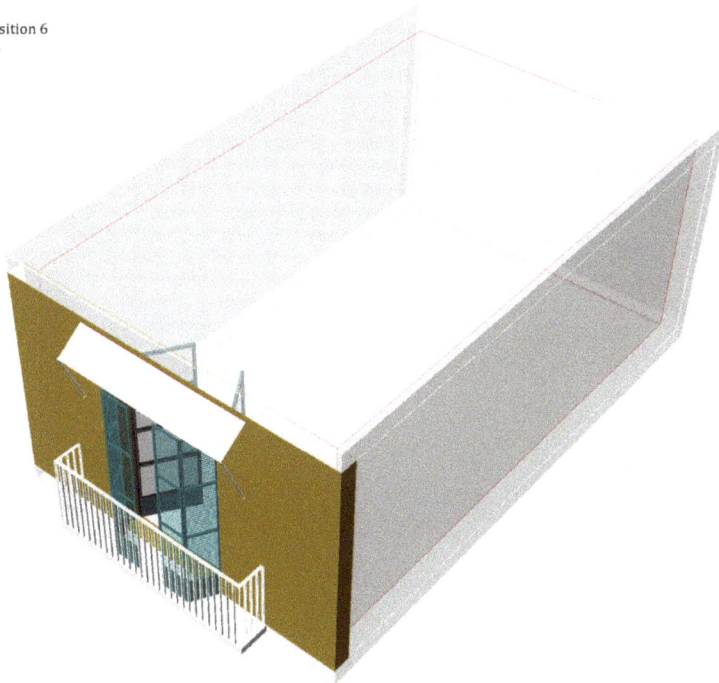

Composition 7

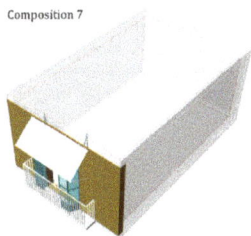

Composition 8

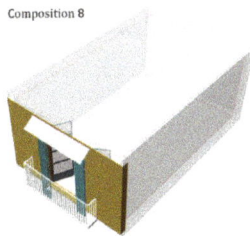

Composition 9

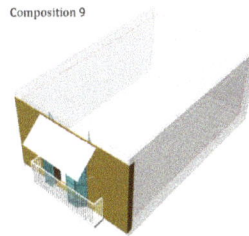

Composition 10

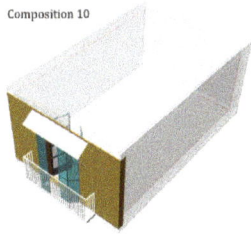

Composition 11

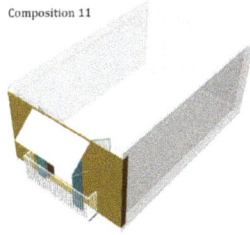

Composition 12

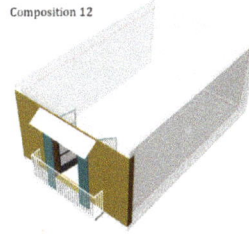

Figure 4. Perspective images of compositions of operable elements found through the described design search procedure. Thermal results related to these configurations are found in Figure 5, and show that only minor perceived temperature variations are found on 21 June. One reason for this is the external blind position, the marquise, combined with the time of the simulation. In Barcelona, the sun sits high during the summer, resulting in a shading effect even when the marquise is only minimally extracted (see results, discussion, and graphs below).

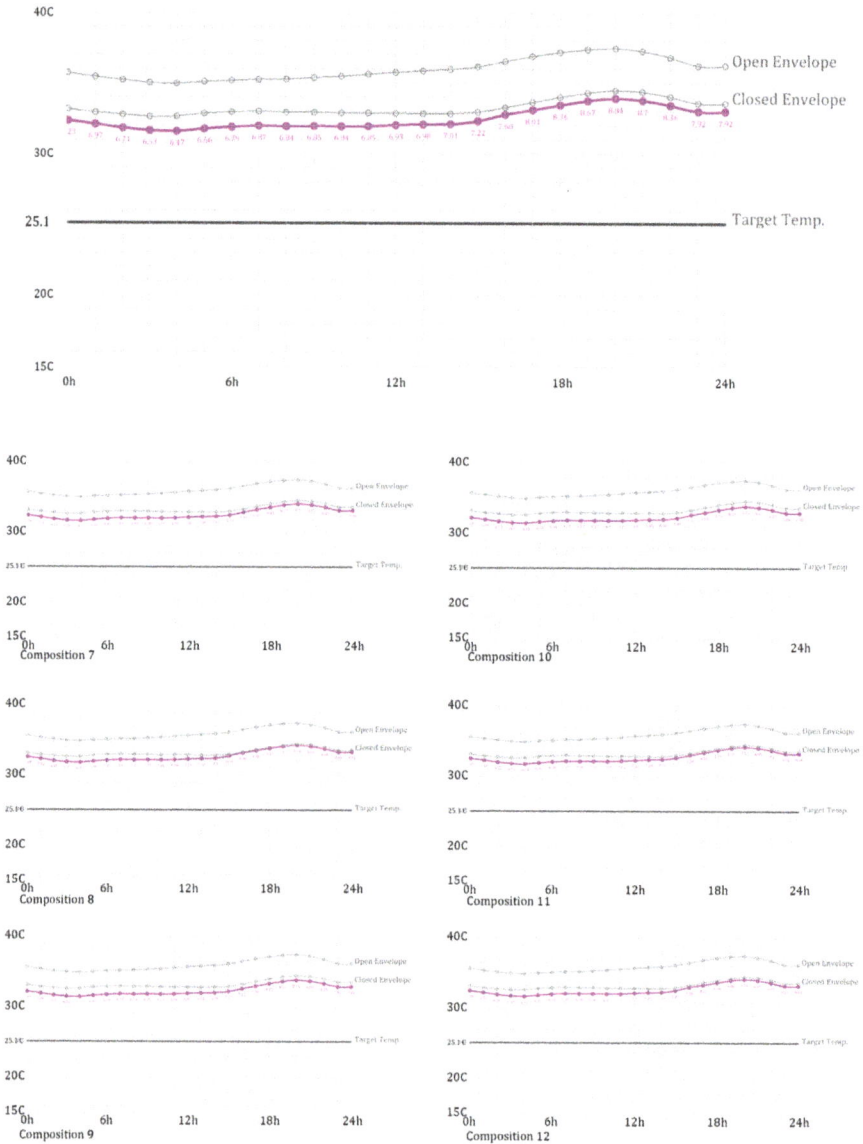

Figure 5. Temperature graphs of temperature deviation for computed compositions 6–12 on 21 June. Related design compositions can be seen in Figure 4.

4. Discussion

From the simulation studies, the compositions of the deep multi-layered envelope enable a significant, but less than expected, temperature regulation. To expand the studies, with the insights gained through this study, a series of thermal effective aspects, such as plant vegetation, thermal storage, cross ventilation, could be added. Previous studies by the author also indicate that coloration has a large impact on the radiation temperature distribution in responsive envelopes, which is a significant part of the operative perceived temperature. Such a study would be more holistic, but may also move focus away from the operable elements, which by human intervention become adaptive systems as part of the microclimatic thermal environment regulated by the human. In this respect, a future design research model will also be expanded to estimate the amount of energy a human is required to use for operating the elements, compared to an automated system with mechanical motor-based actuation of the elements. From this, an exergy-to-thermal performance ratio can be shown, identifying another performance dimension of the vernacular adaptive envelope compared to contemporary mechanized building envelopes.

Post-occupancy thermal data and thermal simulation studies as part of the design process from existing architectures are not compared to the results in this study. Such a study would increase the thermal performance comparison between mechanical, material, and human-based actuation, but also require a substantial insight into the projects compared, as the operational performance of each approach would be needed in high detail to make valid comparative studies. Even in the comprehensive IEA report, descriptions and discussions kept at the strategy and qualitative formulation level. This means a direct comparison of numerical performances between approaches and studies currently are not possible.

From a responsive architecture approach perspective, hybrid systems, combining human, material and mechanical-motor sensing and actuation could lead to novel performances by utilizing the intrinsic-direct capacities and exergy-based approaches of materials and humans. In addition, this would allow the possibility of responsive control without the presence of a human by computational controlled motor systems to create responsive systems, which act prior to thermal deviation events, such as automated windows to allow programmed night cooling counter-balancing temperature rise the following day. Motor-human responsive systems can be seen in the industry [29], and motor-material responsive systems can be seen in academia [30], however, to the author's knowledge, the integration of a direct human-material-motor adaptive system is yet to be developed and studied.

5. Conclusions

The study concludes that adopting, understanding and adapting vernacular architectures, with direct occupancy engagement, towards future subjective thermal adaptive architectures holds two possibilities, for both the design of elements and design of possible compositions of elements. These conclusions are based upon the simulation results presented above and the registrations and following modeling of the many operable elements in proximity to the occupant.

1. The vernacular responsive envelope, as presented and studied, can reduce undesired temperature deviation between the natural thermal environment and the desired thermal environment by the analyzed operable elements. This is particularly true when the thermal sensation is measured against the thermal adaptive comfort model.
2. The vernacular responsive envelope, enable direct occupant adaptation as an exergy-based active agent for subjective thermal regulation, where the building envelope is understood and operated as a thermo-active layer in similarity to actively adjusting clothing levels, in contrast to mechanical-motor based approaches and material-based approaches.

Funding: This research received no external funding.

Conflicts of Interest: The authors declare no conflict of interest.

References

1. Parsons, K. *Human Thermal Environments*, 3rd ed.; CRC Press Tayloy & Francis Group: Boca Raton, FL, USA, 2002.
2. IPCC. Summary for policymakers. In *Climate Chage 2014: Impacts, Adaptation and Vulnerability. Part A: Global and Sectoral Aspects*; Cambridge University Press: Cambridge, UK, 2014.

3. Khotbehsara, E.M.; Nasab, S.N. Porch and balcony as sustainable architecture factors in vernacular houses of west of Guilan: Case studies in Khotbehara, Iran. *World Rural Obs.* **2016**, *8*, 48–56.
4. Motealleha, P.; Zolfagharia, M.; Parsaeeb, M. Investigating climate responsive solutions in vernacular architecture of Bushehr city. *HBRC J.* **2018**, *14*, 215–223. [CrossRef]
5. Bodacha, S.; Langa, W.; Hamhaberc, J. Climate responsive building design strategies of vernacular architecture in Nepal. *Energy Build.* **2014**, *81*, 227–242. [CrossRef]
6. Moe, K. *Convergence: An Architectural Agenda For Energy*; Routledge: New York, NY, USA, 2013.
7. Arens, E.; Hoyt, T.; Zhou, X.; Huang, L.; Zhang, H.; Schiavon, S. Modeling the comfort effects of short-wave solar radiation indoors. *Build. Environ.* **2015**, *88*, 3–9. [CrossRef]
8. Atmaca, I.; Kaynakli, O.; Yigit, A. Effects of radiant temperature on thermal comfort. *Build. Environ.* **2007**, *42*, 3210–3220. [CrossRef]
9. La Gennusa, M.; Nucara, A.; Rizzo, G.; Scaccianoce, G. The calculation of the mean radiant temperature of a subject exposed to the solar radiation—A generalised algorithm. *Build. Environ.* **2005**, *40*, 367–375.
10. Loonen, R.C.G.M.; Trčka, M.; Cóstola, D.; Hensen, J.L.M. Climate adaptive building shells: State-of-the-art and future challenges. *Renew. Sustain. Energy Rev.* **2013**, *25*, 483–493. [CrossRef]
11. Fox, M.; Kemp, M. *Interactive Architecture*; Princeton Architectural Press: Princeton, NJ, USA, 2009.
12. Aschehoug, Ø. *Annex 44: Integrating Environmentally Responsive Elements in Buildings*; State-Of-The-Art-Report; Aalborg University: Aalborg, Denmark, 2009; Volume 1.
13. Perino, M. *Annex 44: Integrating Environmentally Responsive Elements in Buildings*; Responsive Elements in Buildings; Aalborg University: Aalborg, Denmark, 2009; Volume 2A.
14. Menges, A.; Reichert, S. Material Capacity: Embedded Responsiveness. *Archit. Des.* **2012**, *82*, 52–59. [CrossRef]
15. Reichert, S.; Schwinn, T.; la Magna, R.; Waimer, F.; Knippers, J.; Menges, A. Fibrous Structures: An Integrative Approach to Design Computation, Simulation and Fabrication for Lightweight, Glass and Carbon Fibre Composite Structures in Architecture based on Biomimetic Design Principles. *CAD Steer. Form* **2014**, *52*, 27–39. [CrossRef]
16. Pasold, A.; Foged, I. Performative Responsive Architecture Powered by Climate. In Proceedings of the 30th Annual Conference of the Association for Computer Aided Design in Architecture (ACADIA), New York, NY, USA, 21–24 October 2010; pp. 1–14.
17. Foged, I.W.; Pasold, A. An Oak Composite Thermal Envelope. In Proceedings of the ICSA 2016 Conference Proceedings, Venice, Italy, 5–8 April 2016.
18. Torio, H.; Schmidt, D. *Annex 49: Low Exergy Systems for High-Performance Buildings and Communities*. 2011. Available online: http://www.iea-ebc.org/Data/publications/EBC_Annex_49_Factsheet.pdf (accessed on 1 April 2019).
19. Shukuya, M. Exergy concept and its application to the built environment. *Build. Environ.* **2009**, *44*, 1545–1550.
20. Moe, K. *Thermally Active Surfaces in Architecture*; Princeton Architectural Press: Princeton, NJ, USA, 2010.
21. Nicol, F.; Humphreys, M.; Roaf, S. *Adaptive Thermal Comfort: Principles and Practice*, 1st ed.; Routledge: New York, NY, USA, 2012.
22. Humphreys, M.; Nicol, F.; Roaf, S. *Adaptive Thermal Comfort: Foundations and Analysis*, 1st ed.; Routledge: New York, NY, USA, 2015.
23. Leatherbarrow, D. Architecture's Unscripted Performance. In *Performative Architecture—Beyond Instrumentality*; Kolarevic, B., Malkawi, A.M., Eds.; Spon Press: London, UK, 2005; pp. 5–20.
24. De Dear, R.; Brager, G.S. Developing an Adaptive Model of Thermal Comfort and Preference. *Indoor Environ. Qual.* **1998**, *104*, 145–167.
25. Carlucci, S.; Bai, L.; de Dear, R.; Yang, L. Review of adaptive thermal comfort models in built environmental regulatory documents. *Build. Environ.* **2018**, *137*, 73–89.
26. Baker, P. *Technical Paper 10: U-Values and Traditional Buildings*. 2011. Available online: http://eprints.sparaochbevara.se/674/ (accessed on 1 April 2019).
27. Fanger, P.O. *Thermal Confort: Analysis and Applications in Environmental Engineering*; McGraw-Hill: New York, NY, USA, 1972.
28. Fanger, P.O. Thermal environment—Human requirements. *Environmentalist* **1986**, *6*, 275–278.

29. Velux. Velux Integra. 2019. Available online: https://www.velux.dk/sectionsharedfolder/til_professionelle/integra-ovenlysvinduer (accessed on 1 April 2019).

30. Foged, I.W.; Pasold, A.; Pelosini, T. A Hybrid Adaptive Composite Based Auxiliary Envelope. In Proceedings of the International Conference on Structures and Architecture 2019 Conference Proceedings, Lisbon, Portugal, 24–29 July 2019.

buildings

MDPI

Article

Transformed Shell Roof Structures as the Main Determinant in Creative Shaping Building Free Forms Sensitive to Man-Made and Natural Environments

Jacek Abramczyk

Department of Architectural Design and Engineering Graphics, Rzeszow University of Technology, Al. Powstańców Warszawy 12, 35-959 Rzeszów, Poland; jacabram@prz.ed.pl; Tel.: +48-795-486-426

Received: 28 February 2019; Accepted: 20 March 2019; Published: 25 March 2019

Abstract: The article presents author's propositions for shaping free forms of buildings sensitive to harmonious incorporation into built or natural environments. Complex folded structures of buildings roofed with regular shell structures are regarded as the most useful in creative shaping the free forms that can easily adapt to various expected environmental conditions. Three more and more sophisticated methods are proposed for creating variously conditioned free form structures. The first method allows the possibility of combining many single free forms into one structure and leaves the designer full freedom in shaping regular or irregular structures. The second, more sophisticated method introduces additional rules supporting the designer's spatial reasoning and intuition in imposing regularity of the shapes of the building structure and its roof shell structure. The third, most sophisticated method introduces additional conditions allowing the optimization of the regular shapes and arrangement of complete shell roof segments on the basis of an arbitrary reference surface and a finite number of straight lines normal to the surface. This original, interdisciplinary study offers new insight into, and knowledge of, unconventional methods for the creative shaping of innovative free forms, where great possibility and significant restrictions result from geometrical and mechanical properties of the materials used. Solving a number of issues in the field of civil engineering, descriptive geometry and architecture is crucial in the process of creating these structures.

Keywords: building free form structure; corrugated shell roof; integrated architectural form; thin-walled open profile; shape transformation; folded sheet

---

1. Introduction

Curved metal shell roofs have been used since the Gothic and became very popular in the Renaissance owing to their attractive architectural forms and stable constructions [1–3]. Glass and laminated glass elements made of reinforced polymers are used as structural members together with metal ones, which diversifies and improves the attractiveness of the architectural forms of buildings [4–6]. Space grids and complete shells are combined into a single internal coherent shell structure to strengthen the shell roofs and improve their stability [7–10].

Open thin-walled steel sheets folded in one direction joined with their longitudinal edges into flat sheeting can be easily transformed into shell forms as a result of assembling them to skew roof directrices [11,12]. The shell shape of the sheeting depends on a mutual position and curvature of the directrices and can be modelled with the help of warped surface [13]. The transformations are effective if freedom of the transversal width increments of each shell fold at its length is ensured to obtain positive static-strength work [14]. Such transformed sheeting is characterized by big mutual displacements of its subsequent folds in the shell, small strain and big deformations of the fold's flanges and webs [14,15] (Figure 1).

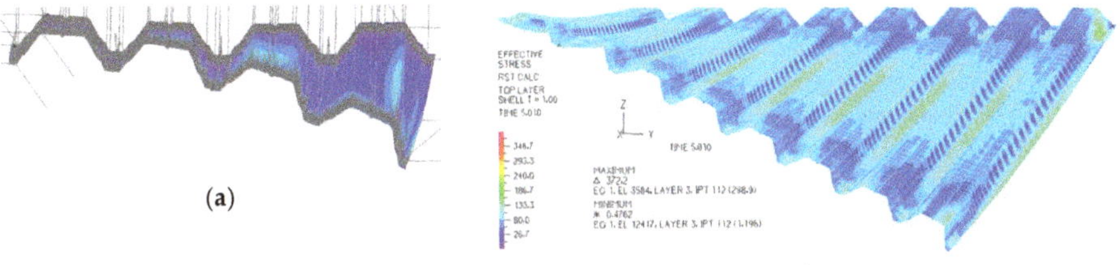

Figure 1. Accurate thin-walled folded computer models: (**a**) a nominally plane folded sheet transformed into a shell shape; (**b**) nominally plane folded sheeting transformed into a shell shape and loaded with a characteristic load.

Because of the above big displacements and effective shape transformations, great freedom in shaping the roof shell forms is achieved by means of two directrices adopted almost freely, so a variety of the architectural free forms of the resultant shell roofs and entire buildings is great [9]. Some important geometrical and mechanical restrictions of the sheet's shape transformations have to be taken into account. The basic one concerns the fact that each effectively transformed fold contracts at its half-length and is stretched at both crosswise ends [16]. Therefore, two or more complete corrugated shell sheets cannot be joined with their crosswise ends, that is perpendicular to the fold's directions, to obtain one resultant smooth shell [17]. They can only be set together with their transverse ends (Figure 2), to obtain an edge roof shell structure with regular edge pattern on its surface [18,19]. For engineering developments, each shell fold can be modelled with a simplified smooth sector of a warped surface [20,21] including hyperbolic paraboloid [22,23]. The sum of all such sectors is a model of a continuous edge structure [13].

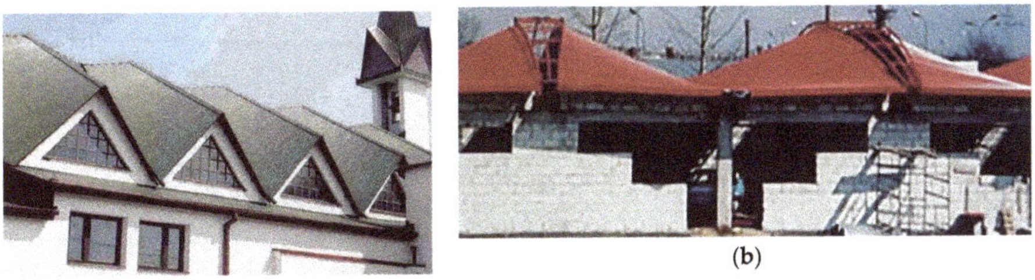

Figure 2. Two shell roof structures characterized by: (**a**) two straight directrices; (**b**) straight and curved directrices.

The considered transformed shells are stiffened with roof directrices transversally in relation to the fold's directions, and additional edge elements in order to maintain the straightness of the border folds in the shell [11–13]. Therefore, the considered structures need respective shapes of stiffened structural systems [7,8,11].

2. Critical Analysis

The use of well-known conventional design methods [6–10,21], known from the traditional courses of theory of structures, in the shaping of shell roof forms is ineffective because it usually results in high values of normal and shear stresses, local buckling and distortion of thin-walled flanges and webs of transformed shell folds. The assembly of the designed shell sheeting into skewed roof directrices is often impossible because of the plasticity of the fold's edges between flanges and webs. Reichhart developed a specific method for calculating the arrangement and the length of the supporting lines of all folds in transformed corrugated shell sheeting [11], but it is effective only for the cases where the fold's longitudinal axes are perpendicular to roof directrices or very close to those [13]. The author significantly improved the Reichhart concept and has proposed an innovative method [13,18], so that the transformation would cause the smallest possible initial stresses on the shell folds resulting from this transformation.

For effective fold transformations, interdependence between the geometrical supporting conditions and the obtained shell forms of a transformed fold in a shell can be used [11]. In these cases, the freedom of the transverse width and height increments of each shell fold forming the transformed sheeting is ensured, and various attractive and innovative shapes of shell roofs and contraction curves of relatively big curvatures on these roofs can be achieved (Figure 3) [13]. If the fold does not have the freedom of transverse width increments due to strong stiffening of its longitudinal edges shared with its adjacent folds, the aforementioned interdependence cannot be used. Neither can it be used if the assembly technique causes additional forces varying the effective widths of the fold ends and their supporting lines.

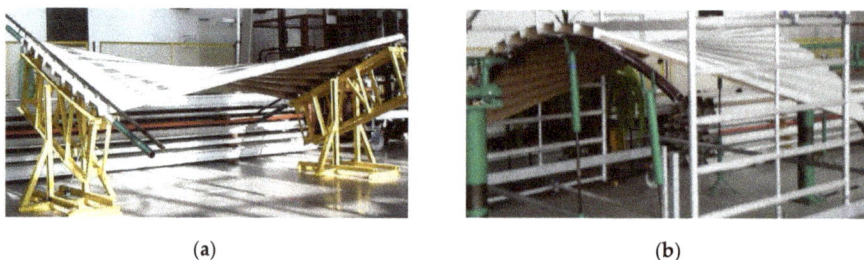

(a) (b)

Figure 3. Experimental transformed corrugated shells supported by: (**a**) straight directrices, (**b**) curved directrices.

In the 1970s, Gergely, Banavalkar and Parker [24] accomplished shape transformations of folded sheets to create shallow right hyperbolic paraboloid roofs and their structures, named "hypars". The very limited shapes of shell roof structures using various configurations of hypars units are also discussed by Bryan and Davies [25]. Right hyperbolic paraboloids are a specific kind of hyperbolic paraboloids whose two rulings belonging to various families of rulings are perpendicular to each other. These two rulings are various lines of contraction of each right hyperbolic paraboloid [13]. Quarters and halves of these central sections are also used and joined together to obtain various shell structures, including hypars [24–26] (Figure 4).

The methods proposed by these authors drastically limit the variety of the designed transformed folded shell forms to central sectors of right hyperbolic paraboloids [26] and their one-fourths [24,25]. Moreover, the models obtained by means of these methods enforce unjustified additional stresses of the folds resulting from the need to adjust the longitudinal axes of the shell folds to the positions of selected rulings of the hyperbolic paraboloid used. The above adjustment of the longitudinal shell fold's axes to these rulings imposes a significant change in the width ΔM (Figure 5) [11] of the transverse fold's ends passing along shell directrix *LM*. These additional forces cause a significant increase in initial stresses and limitation of the searched surfaces to shallow hypars.

Simple shell structures composed of a few corrugated shells have been used in different architectural configurations, most often as shells supported by stiff constructions based on very few columns [27,28]. Shell structures are used for achieving: (a) large spans; (b) greater architectural attractiveness; and (c) skylights letting sunlight into the building interior [29,30].

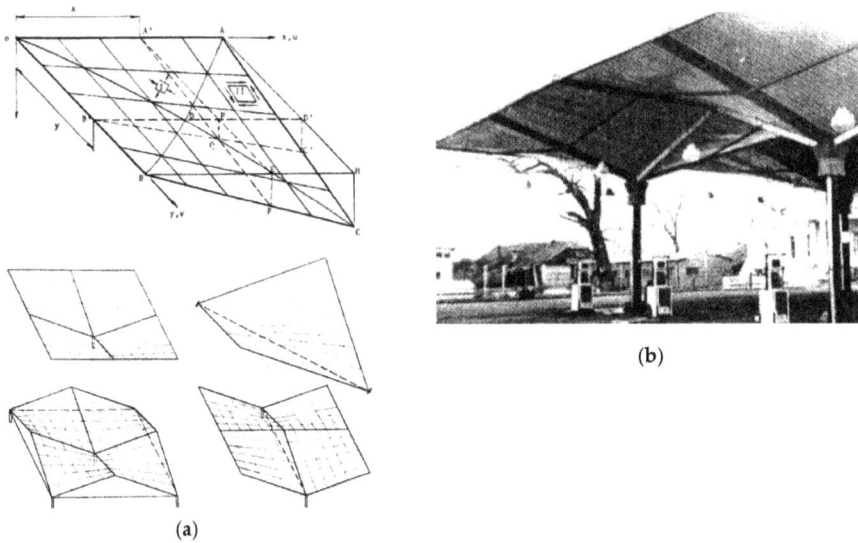

Figure 4. Roof shell structures (**a**) geometrical models; (**b**) erected construction.

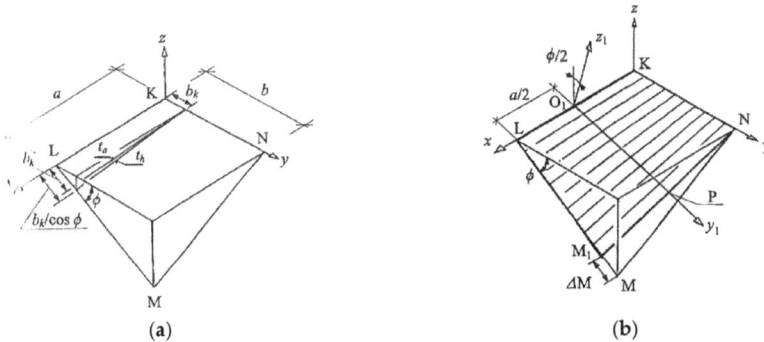

Figure 5. Adjustment of the longitudinal shell fold's axes to the selected rulings of one-fourth of the right hyperbolic paraboloid forcing significant change in the width ΔM of the transverse fold's ends passing along shell directrix *LM*: (**a**) forced change of transverse edge of individual transformed sheet, (**b**) forced change of transverse edge of complete transformed folded shell.

There are very few methods for the geometrical shaping of folded steel roofs transformed into shell forms. Among them, only the Reichhart method allows complete shells different from the central sectors of right hyperbolic paraboloids to be obtained [11]. In the 1990s, Reichhart started to shape corrugated steel sheeting for shell roofing, where all folds underwent big transformations into shell shapes. An additional advantage of the Reichhart method is that the initial stresses induced by the shape transformations are the smallest possible. Reichhart called such transformations free deformations, because they assure freedom of the transversal width and height increments of all folds in the transformed shell. In this way, the initial fold's effort is reduced to a possibly low level.

Reichhart arranged the complete corrugated shells on horizontal or oblique planes [11] as continuous ribbed structures (Figure 2). He developed a simple method for geometrical and strength shaping of the transformed shell roofs. He designed corrugated shell sheeting supported by very stiff frameworks or planar girders with additional intermediate members and roof bracings. [20].

The transverse ends of transformed folds cannot be extended to the positions predicted by conventional methods, because this action causes a radical increase in stresses of the deformed thin-walled profiles [11,13,14]. The fold's fixing points along all roof directrices must be precisely calculated either by the Reichhart method [11], if the longitudinal fold's axes are close to perpendicular to the directrices, or by the author's method [13,18]. For these calculations based on precisely calculated supporting conditions, diversified for the subsequent folds in a roof shell, and the stiffness of these folds resulting from their geometrical and mechanical properties can be applied.

Unfortunately, the Reichhart method is correct only when the longitudinal axes of the transformed shell folds are perpendicular to the roof directrices, or very close to those, and the algebraic equations of these directrices are of the second order at most. In other cases, the method leads to serious errors, as demonstrated by the author [13]. These errors result from the lack of conditions providing similar values of stress at both transverse ends of the same fold. The visible result of different stress values at both transverse ends of the same shell fold is that the transverse contraction of the fold does not pass halfway along its length, on the contrary, it is shifted closer to one of these ends. The condition defined by the author is employed in his innovative method of shaping individual roof shells (monograph) [13] and implemented in the application he developed [18] in the Rhino/Grasshopper program used for parametric modeling of engineering objects.

The author started with experimental tests [13] and computer analyses [14] on static and strength work of folded sheets transformed into various shell forms and structural systems dedicated to supporting the transformed complete roof shells and their complex structures. This issue goes beyond the scope of the article. As Reichhart incorrectly accepted each shell fold as prismatic beams not cooperating with each other and having linear geometrical and mechanical characteristics, the author began his work with a preliminary understanding of the geometrical and static-strength characteristics of thin-walled folded shells. He analyzed the possibilities of modeling the transformed steel sheeting with accurate, thin-walled, folded computer models created in the ADINA program used for advanced dynamic incremental nonlinear analyses [15] (Figure 1). In order to accurately configure his computational models, he intends to perform experimental tests in the near future on the innovative experimental stand of his design at a laboratory hall (Figure 3).

The present article concerns geometrical shaping the building's free forms roofed with folded steel sheeting transformed into various shell forms [31]. Therefore, the possibilities of shaping complex free forms, that is free form structures composed of several single free forms roofed with separate individual transformed shells, are analyzed. The justification for creating roof structures composed of several shell segments results from the geometrical and mechanical properties of the transformed folded sheets employed. The shell folds are twisted around their longitudinal axes or twisted and bent transversely to these axes. That is the reason why the transverse ends of these folds expand and their middles contract as the degree of the transformation increases [13].

Therefore, it is not possible to combine two transformed corrugated shells along their transverse edges into one smooth shell (Figures 2 and 4) [32]. It is possible, however, to join both shells with transverse edges, so that there is an edge between them. The edge disturbs the smoothness of the resultant shell that becomes a structure of two shells. Most often, roof directrices separate the adjacent shell segments in the roof structure, or the adjacent shell segments are separated by additional roof or wall areas that let sunlight into the building's interior.

The main goals of combining complete transformed shells in the roof structure include: increasing the span of the roof and entire building, integrating the roof and façade forms, increasing the visual attractiveness of the entire building free form and making it sensitive to the natural or built environments. The concept most commonly used in the shaping of transformed folded shell structures is the combination of central sections of right hyperbolic paraboloids, their halves or quadrants in various configurations along their common edges (Figure 2a) [25,26].

The variety of shell structures constructed in this way is minimal. Reichhart's actions are also limited to structures composed of several identical central segments of right hyperbolic paraboloids additionally arranged on the same plane (Figure 2) [9].

The author presented wide possibilities of shaping free form structures composed of many individual free forms [13,17]. He developed the concept and coherent rules for creating such complex structures covered with plane-walled folded elevations and multi-segment transformed shell roof structures [18,19]. The developed algorithms allowed a radical increase in the variety of shapes of these forms [19,22,32,33]. Based on these algorithms, the author developed three methods presented in this work. These methods differ in quality and serve to obtain mutually different and specific goals. Each of these methods is aimed at creating building structures of very specific forms, in a convenient and relatively simple way. Therefore, in the author's opinion, only a qualitative comparison of these methods is justified.

The methods and examples presented in the sections that follow describe step by step and define objects, actions and algorithms used to solve increasingly complex issues of shaping internally consistent forms of building structures sensitive to the natural or built environments. The structure of the present article has been adopted so as to discuss step by step the specifics of the search for more and more sophisticated forms of building free forms roofed with transformed corrugated steel shell structures.

The designer may have to face, and cope with, some problems that arise from using unconventional methods for shaping general architectural forms of buildings roofed with transformed folded steel sheets and striving for relatively simple implementation of the designed innovative forms. The main task is to achieve geometrical, architectural and structural cohesion of all elements of each free form building, its shell roof in particular [15,32]. This aim can be accomplished by creating a parametric description of such building free forms and their specific structural systems based on the geometrical and mechanical characteristics of the transformed sheeting [14]. The proposed methods contain geometrical descriptions and algorithms that can be employed in the creating of parametric description of the free form structures covered with plane-walled folded elevations and complex transformed steel roof shell structures and writing parametric computer applications assisting the designer in the engineering developments.

Prokopska and the author continue the problems initiated by Reichhart. They propose a method of geometrical integration of each shell roof form with plane and oblique walls to obtain innovative, attractive and multi-variant architectural forms considered as morphological systems of buildings [18]. Some main principles of shaping complete and compound innovative free forms are the result of the cooperation between Prokopska and the author [32].

On the basis of these principles the author invented two methods for parametric shaping of the complete architectural free forms and their complex structures covered with transformed folded shell steel sheeting [18,22,31]. He assumed that the great freedom in shaping diversified transformed shell forms for roofing, resulting from great freedom in adopting shapes and mutual positions of roof directrices, can be used to integrate the entire building free form and make the form very sensitive to the natural or built environments [32]. Consequently, to achieve more consistent and sensitive architectural free forms, he decided to fold and incline elevation walls to the vertical depending on the shape of the shell roof and entire building. He noticed that the interdependence between the efficiency of the roof sheeting transformation and the location of its contraction along the length of each shell roof fold greatly enhances the attractiveness of the entire form and the integrity of the shapes of the roof and elevation [17].

Prokopska conducted multivariate interdisciplinary analyses of some consistent morphological systems that can be designed in harmony with the natural or man-made environments. Her research involves many interdisciplinary topics needed to develop experience in shaping various attractive architectural free forms [34,35]. Some of the proposed structural systems [36,37] can be modified and employed in the discussed building free forms [17].

3. Aims and Scope of the Article

The aim is to present new possibilities for the geometrical shaping of the free form structures of buildings roofed with many transformed shell segments, using three methods that differ in the complexity of algorithms, the purposes they can be used for, and regularities. The methods, presented in such a proposed order, are increasingly sophisticated in the creative search for coherent forms of the complex free buildings sensitive to the natural or built environments. They allow for obtaining these structures that differ in qualitative rather than quantitative properties of free forms, whose creation is discussed in the article.

Especially in the third method—the most complex and sophisticated method—a regular polyhedral network composed of many regular specific tetrahedrons is defined, such that the position of their side edges is optimized in relation to a finite number of selected straight lines normal to almost any double curved auxiliary regular surface, called the reference surface. The proposed rules, objects and activities ensure the regularity of the roof structure, integration of the structure with the folded façade form and allow the free form building to be adapted to the natural or built environments.

The dimensions of the roof shell and façade walls, and the inclination of their characteristic edges can be freely and creatively shaped as well as modified in the consequent steeps of the algorithms proposed by these methods, according to the expected engineering developments.

4. The Concept and the Range of the Article

The article proposes three methods for shaping complex free form buildings roofed with structures of corrugated shells made up of nominally flat folded sheets connected to each other by longitudinal edges, and transformed into spatial forms. The presentation of the methods on specific examples began with the simplest formulation based on a few very simple rules. The next two methods are increasingly complex and serve to achieve different sophisticated goals. Therefore, only a qualitative comparison of these three methods can be made. Comparison of the quantitative results achieved with the methods and included those in the specific examples presented is not justified. The quantitative comparison of these methods with the methods of other authors mentioned in the previous sections also seems unjustified.

The first method formulates the basic conditions that must be met by single free forms Σ covered with single transformed shells Ω (Figure 6), so that they can be combined into a structure in a simple way. The combined structure has a free form, and is roofed with a structure composed of several shell segments. The purpose of creating a roof structure is to increase the span of a complex building form in relation to the span of a single form.

The basic action of the first method is to create a model of roof eaves of each complete free form. The model is a closed spatial quadrangle B_{ev}, whose geometrical properties depend on the form of rectangular, nominally flat sheets folded in one direction and transformed into a shell form. Points B_i are four vertices of B_{ev}.

Since the folded sheets are rectangular, the angles between two adjacent sides having shared ends at corners B_j (j = 1–4) of the spatial quadrangle B_{ev} are very close to right angles, and the lengths of each pair of the opposite sides of the quadrangle are equal to each other or differ very little. Two opposite sides of the aforementioned quadrangle, corresponding to the transverse edges of a transformed shell are almost equal to each other. Furthermore, if the transverse edges of the shell are not obliquely cut [13], the lengths of all folds of the shell are identical, so the second pair of the opposite sides of this quadrangle is formed from two skew straight sections of equal length. The transverse fold ends are often cut obliquely to adjust the transverse edges of the shell to the direction of the roof directrices. However, the cuts are only minor and cause little variation in the fold's length, followed by a slight difference in the lengths of the opposite sides and the measures of the corner angles of the aforementioned spatial quadrangle.

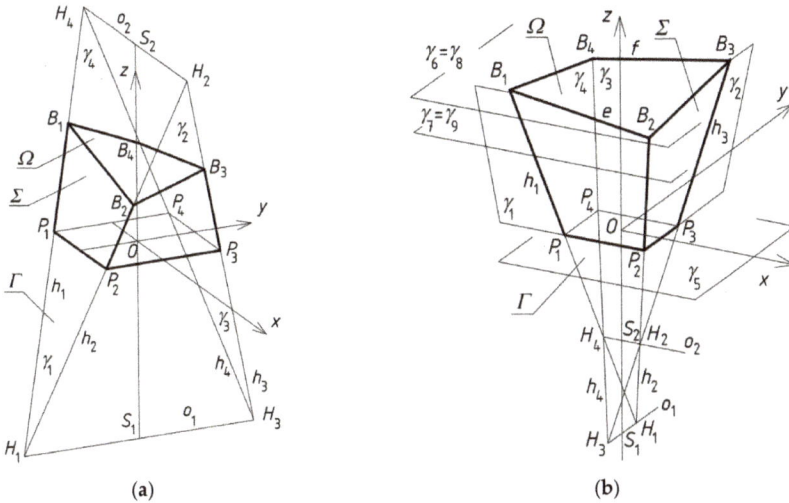

Figure 6. Two simplified models Σ of a free form building roofed with transformed shells Ω: (**a**) straight directrices; (**b**) curved directrices e and f.

As all folds of each transformed roof shell are almost always twisted along the longitudinal axes, expanded at the transverse ends and contracted at half-length, each quadrangle cannot be flat; on the contrary, it must be spatial. Each pair of its two opposite sides is created as two skewed lines. The lengths of these sides must be precisely calculated based on the border conditions adopted for the roof's corrugated shell. The conditions are determined mainly by the shape and mutual position of the roof directrices [14,18].

For the above models and in the initial example presented in Section 5, the directrices are adopted as two straight segments $e = B_1B_2$ and $f = B_3B_4$ in order to obtain a relatively simple description of the first method. The B_{ev} spatial quadrilateral is made up of two pairs of skew straight sections. One pair of these sections is formed of the e and f directrices corresponding to the shell's edges running transversely to the shell fold directions. The second pair of the opposite straight sections corresponds to the longitudinal edges of the shell belonging to the extreme folds.

Since the directrices are skewed straight lines, the calculated supporting conditions of the subsequent folds in a shell, mainly affecting the fold's twist, are varied. As a result, the subsequent folds have different lengths of their supporting lines, twists and lengths. The author attempts to use quadrangles B_{ev} whose shapes are symmetrical towards an axis skew with respect to each side of quadrangle B_{ev} in order to obtain pairs of its opposite edges of equal length and congruent apex angles. Such an operation leads to a symmetrical complete roof shell, identical inclination of each shell fold to both directrices and identical supporting conditions of the fold at both ends.

In order to build a very simple free form whose roof eaves are spatial quadrangle B_{ev} characterized by the aforementioned properties, the definition of four planes γ_i modeling four façade walls of this form is most convenient (Figure 6). In these planes, four straight sections of B_{ev} with common vertices are defined. On the basis of these elements, the author defined a simplified model of a single free form, and called it the reference tetrahedron Γ [13]. He distinguished three basic types of the reference tetrahedrons. Two of them are presented in Figure 6.

He also drew attention to the following basic geometrical properties of each reference tetrahedron Γ_i of the free form structure Γ, where i indicates the number of all reference tetrahedrons used for creating free form structure Σ. For each reference tetrahedron, two of the six edges formed as a result of the intersection of the above four planes are called axes o_1 and o_2 (Figure 6), and the other are side edges h_j (j = 1–4). The axes are the intersecting lines of two opposite planes of the tetrahedron. The side edges are the intersection of the adjacent planes of tetrahedron Γ_i. The axes intersect with the side edges at points H_j called the vertices of Γ_i. Vertices B_j of each quadrangle B_{evi} are defined at side edges h_j at adequate distances from the respective vertices H_j. Four sides of B_{ev} are created on the basis of four vertices B_j.

In order to obtain spatial network Γ composed of many complete reference tetrahedrons Γ_i, the reference tetrahedrons are arranged so that one wall of each two neighboring tetrahedrons is contained in a common plane. A detailed description of the relations between the axes, edges and vertices of neighboring reference tetrahedrons Γ_i in Γ is presented in Section 5 concerning the first method proposed by the author.

The method enables the full recognition of the possibilities of combining the reference tetrahedrons into one spatial reference network Γ whose unconventional and innovative form is determined by the specific properties of these tetrahedrons Γ_i. The method leaves the designer complete freedom to give the shape of the spatial reference network Γ and the roof shell structure Σ. As a result, the network may be regular or not. The arrangement of the shell segments in the three-dimensional space may be regular or not. The forms of the individual shells may be regular and similar to each other or not, depending on the geometrical properties of the subsequently adopted complete free forms.

As reference network Γ is used to create regular free form building structures roofed with regular shell roof structures, it is necessary to define additional rules supporting the designer's spatial reasoning and intuition. Therefore, the author developed the second method of shaping the aforementioned structures roofed with transformed curved shells. In order to impose the regularity of a complex free form and its roof shell structure, a so-called reference surface is introduced into the method. According to the method, the shapes and arrangement of the reference tetrahedrons in the three-dimensional space should be determined on the basis of the reference surface.

Using this method, it is possible to exploit specific geometrical properties of the reference surface, such as its planes of symmetry, so that the symmetry axes of the largest possible number of reference tetrahedrons Γ_i are contained in these planes. Moreover, relatively simple operations are possible to obtain the symmetry of quadrangles B_{evi} contained in Γ_i. In this method, however, reference tetrahedrons Γ_i are arranged on the basis of the reference surface in an intuitive manner to a large extent, whereby their construction is related to the curvatures of the reference surface only to a small extent. This may lead to unjustified differences in the shapes and irregular distribution of the reference tetrahedrons in relation to the reference surface, so certain additional rules are needed. Therefore, the author decided to employ some planes normal to the reference surface for selected planes of the reference tetrahedrons.

In order to prevent inefficient and irregular forms of the reference network and shell roof structure, the author created the third method and introduced another condition allowing him to optimize the regular shapes and arrangement of the reference tetrahedrons in relation to the arbitrary reference surface. He imposed a constraint that the directions of all side edges of the reference network should be close to the directions of the adopted normals of the reference surface at the points of the intersection of the above side edges and the reference surface within the adopted optimization accuracy. Such operations, undertaken by the author, result in the fact that the projections of the single roof shell segments on a reference surface, in the directions compatible with the directions of selected straight lines normal to this surface, do not overlap each other. Instead, they form a continuous two-dimensional surface. In other words, the projections are not disjoint and do not create discontinuous areas.

The third method allows an optimization of the positions and directions of the side edges $h_{i,j}$ of reference tetrahedrons $\Gamma_{i,j}$ relative to the arbitrary reference surface. The optimization is carried out using the aforementioned straight lines $n_{i,j}$ normal to the reference surface. The optimizing condition is the allowable size of the deviation of all edge sides from the corresponding normals to the reference surface.

The basic difficulty in maintaining the above optimizing condition is the fact that two subsequent straight lines $n_{i,j}$ normal to the reference surface are skew lines, in contrast to each pair of neighboring edges $h_{i,j}$ of each reference tetrahedron $\Gamma_{i,j}$, which intersect each other at the appropriate vertex $H_{i,j}$ of $\Gamma_{i,j}$. Therefore the perfect approximation or replacement of $n_{i,j}$ with $h_{i,j}$ is impossible, and the optimization of the side edges $h_{i,j}$ in relation to the reference surface is needed. In the third method, the author therefore included an innovative way of looking for a regular spatial network composed of regular reference tetrahedrons whose side edges $h_{i,j}$, intersecting each other at vertices $H_{i,j}$ of reference tetrahedrons $\Gamma_{i,j}$, are optimized and defined on the basis of not crossing each other's normals with regard to the reference surface.

5. Structures as Compositions of Many Regular Free Forms

The following steps, activities and objects have been identified in the algorithm of the first developed method for shaping complex free form structures. In the first step, some actions are undertaken to build the general form of a single free form with the help of reference tetrahedron Γ_1, whose four walls model four façade walls of the designed free form.

In the second step, the roof of a single free form Σ_1 is modeled as a sector of a smooth regular warped surface. The roof is made of nominally flat corrugated sheets transformed effectively into a shell shape; that is, the shell shape resulting from this transformation should contract at half-length of each fold, transversally to the fold directions, which positively affects the static-strength work of the folds in the shell.

In the third step of the algorithm, a method for determining the positions of several individual forms and combining these forms into a structure sensitive to the predicted natural or man-made environments is carried out. The complex building free form is, therefore, the sum of several individual free forms appropriately set together with the common façade walls. The roof of the building structure created this way is a shell roof structure composed of several smooth shell segments.

In the last step of the algorithm, modification of the forms of the roof and façade of the previously achieved structure is possible. This modification is based on displacements of selected roof edges or façades, in the planes of the reference tetrahedrons employed. The purpose of this modification is to make the building structure more sensitive to the built or natural environments. Complete reference tetrahedron Γ_1 (Figure 7a) is formed by means of four vertices: $H_{1,1}$, $H_{2,1}$, $H_{3,1}$, $H_{4,1}$. To determine the positions of the above four vertices, two skewed straight lines $o_{1,1}$ and $o_{2,1}$ located in distance d_{n1} and perpendicular to each other are assumed. Middle point $S_{2,1}$ of $H_{2,1}H_{4,1}$, is lain in the distance d_{p1} from the origin O of the orthogonal coordinate system $[x,y,z]$. A straight line z perpendicular to $o_{1,1}$ and $o_{2,1}$ intersects these lines at points $S_{1,1}$, and $S_{2,1}$. Finally, the vertices $H_{3,1}$, $H_{1,1}$ are measured along $o_{1,1}$ in the distances $d_{3,1}$ and $d_{1,1}$ from $S_{1,1}$, and the vertices $H_{4,1}$, $H_{2,1}$ are measured along $o_{2,1}$ in the distances $d_{4,1}$ and $d_{2,1}$ from $S_{2,1}$.

Lines $o_{1,1}$, and $o_{2,1}$ are called the axes of Γ_1. However, straight lines: $h_{1,1}(H_{1,1}, H_{4,1})$, $h_{2,1}(H_{1,1}, H_{2,1})$, $h_{3,1}(H_{2,1}, H_{3,1})$, $h_{4,1}(H_{3,1}, H_{4,1})$ are said to be the side edges of Γ_1. In order to obtain the vertices: $P_{1,1}$, $P_{2,1}$, $P_{3,1}$, $P_{4,1}$ of a planar base of $\Gamma_{1,1}$, plane $(x, y) \perp z$ is passed through point O located in the distance d_{p1} from $S_{2,1}$, (Figure 7a). Vertices $B_{1,1}$, $B_{2,1}$, $B_{3,1}$, $B_{4,1}$ of the free form Σ_1 are constructed on $h_{1,1}$, $h_{2,1}$, $h_{3,1}$, $h_{4,1}$ in the distances $d_{h1,1}$, $d_{h2,1}$, $d_{h3,1}$, $d_{h4,1}$ from the above points $P_{i,1}$ ($i = 1$–4). Values of all above input data are presented in Table 1. The above four values are adopted so that the eaves of the resultant Σ_1 are characterized by two pairs of opposite segments of equal length to obtain a central sector of an oblique hyperbolic paraboloid [19].

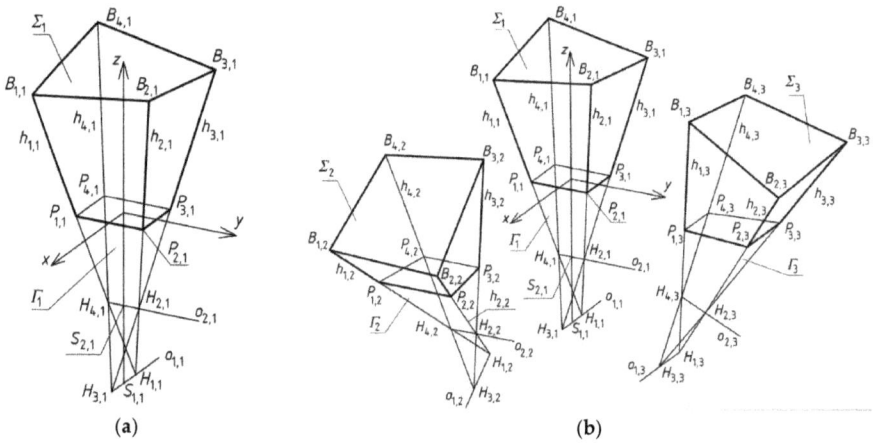

Figure 7. (a) Properties of a reference tetrahedron; (b) Composition of three reference tetrahedrons located orthogonally in reference structure.

Table 1. Parameters providing the parametric characteristics of reference tetrahedron Γ_1 and free form structure Σ_1.

Parameter	Value
$d_{1,1} = d_{3,1}$	3310.6
$d_{2,1} = d_{4,1}$	2231.8
d_{n1}	10,500.0
d_{p1}	12,500.0
$d_{h2,1}$	21,827.3
$d_{h3,1}$	18,827.3
$d_{h1,1}$	18,827.3
$d_{h4,1}$	21,827.3

To determine the subsequent reference tetrahedrons of reference tetrahedral structure Γ being sought, one should create nine complete free forms Γ_j ($j = 1$–9) of Γ. In order to create reference tetrahedron Γ_2, four vertices $H_{i,2}$ ($i = 1$–4) (Figure 7b) should be adopted as previously the vertices of Γ_2. In order to create free form Σ_2, vertices $B_{i,1}$ and $P_{i,1}$ ($i = 1$–4) should be adopted on the basis of the appropriate parameters as previously.

The same action must be performed to obtain reference tetrahedron Γ_3 and free form Σ_3 (Figure 7b). However, for the designed reference polyhedral structure Γ, other activities are undertaken to simplify the assembly of reference tetrahedrons Γ_j ($j = 1$–9) into one structure Γ and complete free forms Σ_j ($j = 1$–9) into one structure Σ.

According to the aforementioned concept, it is taken that $H_{3,2} = H_{1,1}$, $H_{2,2} \in h_{1,2}$, $H_{4,2} \in h_{1,1}$ (Figure 8). It is assumed that $H_{2,2} = H_{2,1}$, $H_{4,2} = H_{4,1}$. To obtain Γ_2 symmetrical to Γ_1 towards plane ξ_2 ($H_{3,2}$, $H_{2,2}$, $H_{4,2}$), vertex $H_{1,2}$ being sought has to be symmetrical to $H_{3,1}$ towards ξ_2.

The transformation related to plane ξ_2 of symmetry is denoted as L_2, so $H_{1,2} = L_2(H_{3,1})$, $B_{1,2} = L_2(B_{4,1})$, $B_{2,2} = L_2(B_{3,1})$, $\Gamma_2 = L_2(\Gamma_1)$. In addition, $B_{3,2} = B_{2,1}$, $B_{4,2} = B_{1,1}$, $P_{3,2} = P_{2,1}$, $P_{4,2} = P_{1,1}$ and $P_{1,2} = (H_{1,2}, H_{4,2}) \cap (x, y)$ and $P_{2,2} = (H_{1,2}, H_{2,2}) \cap (x, y)$.

Reference tetrahedron Γ_3 is created in the same way as Γ_2 so that plane ξ_3 ($H_{4,3} = H_{2,1}$, $H_{3,3} = H_{3,1}$, $H_{1,3} = H_{1,1}$) is used for transformation L_3. Thus, $B_{3,3} = L_3(B_{4,1})$, $B_{2,3} = L_3(B_{1,1})$, $\Gamma_3 = L_3(\Gamma_1)$.

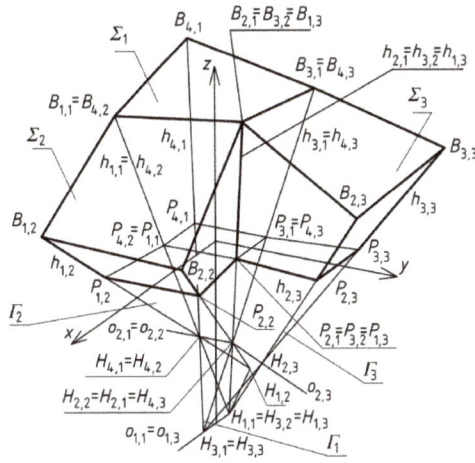

Figure 8. Three reference tetrahedrons: Γ_1, Γ_2 and Γ_3 located orthogonally in the polyhedral reference structure.

Γ_2 and Γ_3 are located orthogonally in Γ. The reference tetrahedron Γ_4 being sought is located diagonally in Γ. The way of creating Γ_4 is different from Γ_2 and Γ_3 because the locations of two its vertices are known: $H_{4,4} = H_{2,1}$, $H_{3,4} = H_{1,1}$ (Figure 9). The searched vertices $H_{1,4}$ and $H_{2,4}$ have to belong to $h_{1,4} = h_{2,2}$ and $h_{3,4} = h_{2,3}$, respectively.

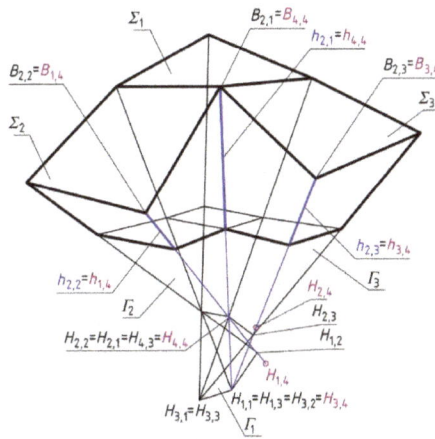

Figure 9. Construction of the forth diagonal reference tetrahedron Γ_4 on the basis of three reference tetrahedrons: Γ_1, Γ_2 and Γ_3 located orthogonally in the polyhedral reference structure.

To obtain Σ_4, vertices $H_{1,4}$ and $H_{2,4}$ are accepted at side edges $h_{1,4}$ and $h_{3,4}$ (Figure 10). It is assumed that $B_{4,4} = B_{2,1}$, $B_{3,4} = B_{2,3}$, $B_{1,4} = B_{2,2}$. The position of $B_{2,4}$ is determined at $h_{2,4}(H_{1,4}, H_{2,4})$. Point $P_{2,4}$ is the intersection of $h_{2,4}$ and the base plane (x,y). The data used in the present example are given in Tables 2 and 3.

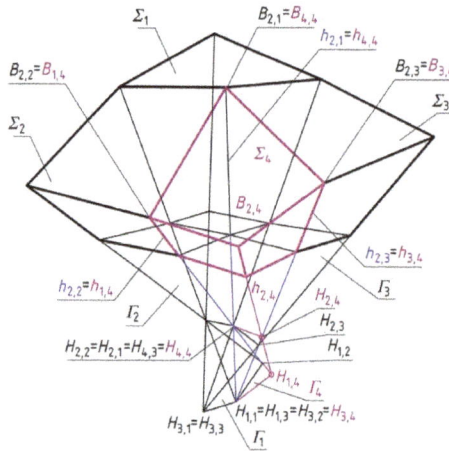

Figure 10. Construction of the forth diagonal reference tetrahedron on the basis of three reference tetrahedrons: Γ_1, Γ_2 and Γ_3 located orthogonally in the polyhedral reference structure.

Table 2. Coordinates of the roof and elevations edges vertices of basic free form Σ.

Vertex	X-Coordinate	Y-Coordinate	Z-Coordinate
$B_{2,1}$	10,373.9	9225.2	20,402.1
$B_{3,1}$	−9489.7	8629.2	17,598.0
$B_{1,1}$	9489.7	−8629.2	17,598.0
$B_{4,1}$	−10,373.9	−9225.2	20,402.1
$B_{2,2}$	25,036.9	8629.2	6711.9
$B_{1,2}$	27,369.5	−9225.2	8501.8
$B_{2,3}$	9489.7	24,395.6	10,578.5
$B_{3,3}$	−10,373.9	26,080.6	12,897.8
$B_{2,4}$	29,067.4	26,269.0	6580.0
$P_{2,1}$	3941.2	4888.7	0.0
$P_{3,1}$	−3941.2	4888.7	0.0
$P_{1,1}$	3941.2	−4888.7	0.0
$P_{4,1}$	−3941.2	−4888.7	0.0
$P_{2,2}$	16,290.0	6394.2	0.0
$P_{1,2}$	16,290.0	−6394.2	0.0
$P_{2,3}$	5457.2	16,710.1	0.0
$P_{3,3}$	−5457.2	16,710.1	0.0
$P_{2,4}$	19,496.9	19,889.3	0.0

Table 3. Coordinates of the vertices of reference structure Γ.

Vertex	X-Coordinate	Y-Coordinate	Z-Coordinate
$H_{1,1}$	−3310.6	0.0	−2300.0
$H_{3,1}$	3310.6	0.0	−2300.0
$H_{2,1}$	0.0	2231.8	−12,500.0
$H_{4,1}$	0.0	−2231.8	−12,500.0
$H_{1,2}$	−8734.4	0.0	−19,202.3
$H_{2,3}$	0.0	6309.5	−14,315.5
$H_{1,4}$	−11,337.6	−665.2	−21,199.8
$H_{2,4}$	470.4	7206.1	−13,081.4

In addition, the following dependences should be adopted: $B_{1,3} = B_{2,1}$, $B_{4,3} = B_{3,1}$, $P_{1,3} = P_{2,1}$, $P_{4,3} = P_{3,1}$, $B_{1,4} = B_{2,2}$, $B_{3,4} = B_{2,3}$, $B_{4,4} = B_{2,1}$, $P_{1,4} = P_{2,2}$, $P_{3,4} = P_{2,3}$, $P_{4,4} = P_{2,1}$, $H_{1,3} = H_{1,1}$, $H_{3,3} = H_{3,1}$,

$H_{4,3} = H_{2,1}$, $H_{3,4} = H_{3,2}$, $H_{4,4} = H_{2,1}$. Other vertices are symmetrical towards plane (x,z) or (y,z). Thus, in order to obtain the entire reference structure Γ and free form structure Σ, four reference polyhedrons Γ_i $(i = 1–4)$ and four free forms Σ_i have to be transformed towards these planes of symmetry (x,z) and (y,z) into new positions of Γ_r and Σ_r $(r = 5–9)$. The final reference structure Γ and free form structure Σ, are shown in Figure 11.

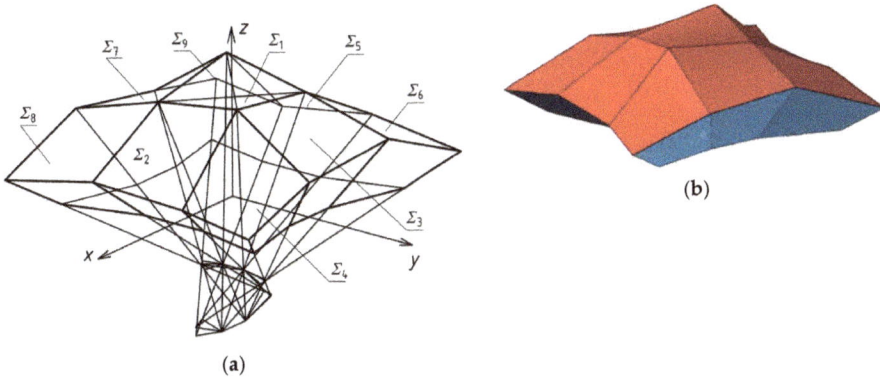

Figure 11. The free form structure created on the basis of the polyhedral reference structure: (**a**) an edge model; (**b**) a shell model.

It is worth stressing that the reference tetrahedrons located orthogonally or diagonally in Γ do not have to be congruent to each other, so the form of the final Γ and Σ can take unsymmetrical shapes. Thus, the possible shapes of Σ may be really free, diversified and sensitive to the built and natural environments.

The free form presented previously is the basis of creating some derivative forms. The derivative free forms are shaped as a result of displacing or rotating some selected side edges of the basic free form in selected planes of its reference structure. The basic form is covered with the continuous shell structure (Figure 11b), whose individual shells are divided by shared edges locally disturbing the smoothness of the structure and forming a regular pattern on the roof. In addition, the elevations have relatively simple shapes.

The first derivative form is constructed as a result of displacing the selected vertices belonging to five roof border quadrangles located diagonally in Γ and distinguished by means of a black thick line in Figure 12. Their selected vertices $B_{i,j}$ are displaced along the relevant side edges $h_{i,j}$ in the distances equal to $d_h = 3000.0$ mm. The values of the coordinates of the points used are presented in Table 4.

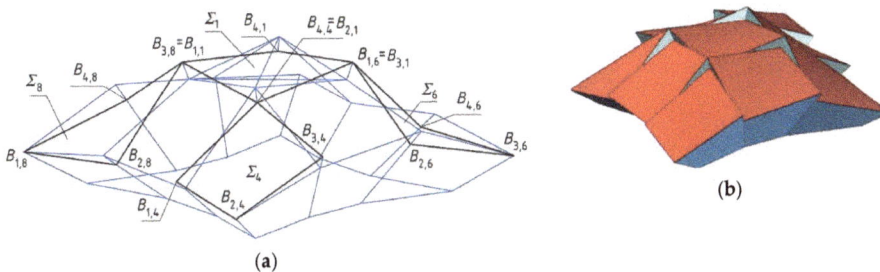

Figure 12. The new transformed discontinuous free form structure created on the basis of the continuous free form structure presented earlier: (**a**) an edge model; (**b**) a shell model.

Table 4. Coordinates of the selected roof vertices of the first derivative form.

Vertex	X-Coordinate	Y-Coordinate	Z-Coordinate
$B_{2,1}$	9489.7	8629.2	17,598.0
$B_{1,1}$	10,373.9	−9225.2	20,402.1
$B_{1,4}$	27,369.5	9225.2	8501.8
$B_{3,4}$	10,373.9	26,080.6	12,897.8

For the aforementioned free form, the parametric shape characteristics can be improved and extended by new shape parameters, for example describing the proportion of the roof discontinuity areas intended for windows to the area of the entire shell roof. Propositions of such additional shape parameters go beyond the scope of the paper.

The second derivative form (Figure 13) is created as a result of: (a) the translations of all vertices of the eaves of Σ_1 discussed earlier, along the relevant side edges of Γ_1 in equal distances d_h; and (b) the translations of the selected elevation side edges along the selected axes of the reference structure in the distances $d_{o1,2} = H_{3,2}H_{1,1} = 4534.1$ mm and $d_{o2,1} = H_{2,3}H_{4,3} = 3126.7$ mm, so that these side edges will be contained in the planes of control compositions Γ_2 and Γ_3. The values of the parameters used are as follows: $B_{3,1}B_{4,4} = B_{3,1}B_{2,1} = d_{h1,1} = 3000.0$ mm. The values of the coordinates of the transformed vertices of the eaves are included in Table 5.

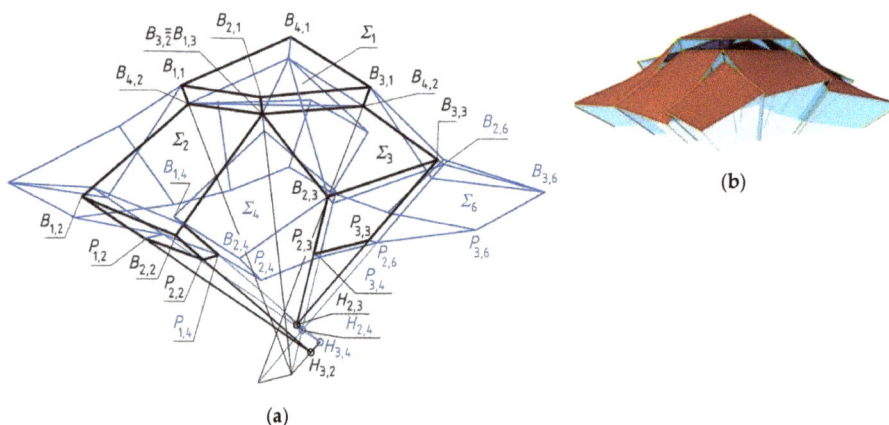

(a)

(b)

Figure 13. The transformed discontinuous free form structure created on the basis of the continuous free form structure presented earlier: (**a**) an edge model; (**b**) a shell model.

Table 5. Coordinates of the roof and elevations vertices of the second derivative form.

Vertex	X-Coordinate	Y-Coordinate	Z-Coordinate
$B_{2,1}$	11,258.0	9821.2	23,206.2
$B_{1,1}$	10,373.9	−9225.2	20,402.1
$B_{2,2}$	26,511.9	8569.2	5334.8
$B_{1,2}$	29,335.6	−9290.8	7501.5
$B_{2,3}$	9553.8	23296.3	11,290.3
$B_{3,3}$	−10,305.4	24728.7	13,261.9
$P_{2,2}$	19,559.6	6792.8	0.0
$P_{1,2}$	19,559.6	−6792.8	0.0
$P_{2,3}$	5249.9	15,093.6	0.0
$P_{3,3}$	−5249.9	15,093.6	0.0

6. Structures Based on Regular Spatial Networks

A new method for shaping the free form structures is presented in the example of nine complete shells located towards a reference ellipsoid ω (Figure 14) is described. Ellipsoid ω is expressed as:

$$\frac{x^2}{a^2} + \frac{y^2}{b^2} + \frac{z^2}{c^2} = 1 \tag{1}$$

where a = 25,000 mm, b = 20,000 mm, c = 13,000 mm.

The orthogonal coordinate system $[x, y, z]$ having its origin at centre O of ω (Figure 14) is adopted. Three basic ellipses w_0, t_0, u_0 are the intersection of ω with planes (y, z), (x, z) and (x, y), respectively. In the examples presented below, the symbols of all created objects, for example reference tetrahedrons and their vertices, have been changed for a more consistent description of creating reference networks.

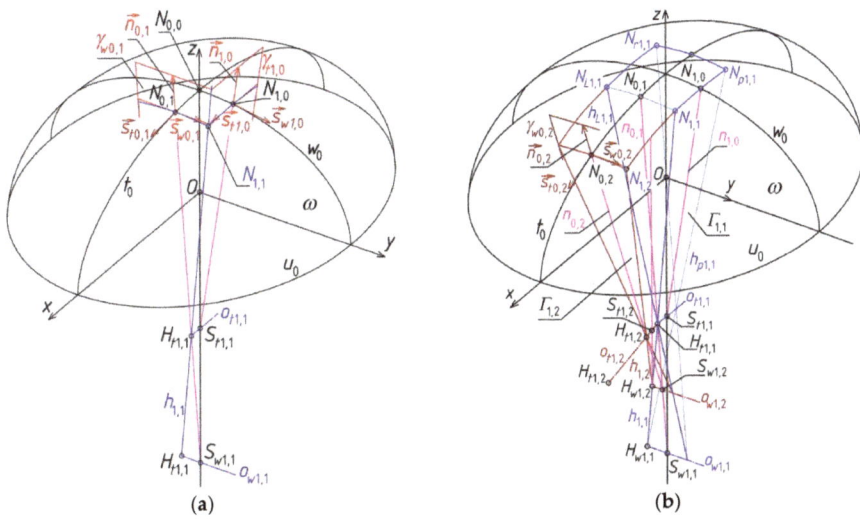

Figure 14. Shaping of: (a) a complete free form with the help of a reference tetrahedron; (b) a free form structure with the help of few reference tetrahedrons.

The planes $\gamma_{w0,1}$ and $\gamma_{t1,0}$ from among four planes of the first reference polyhedron $\Gamma_{1,1}$ are presented in Figure 14. The positions of these four planes of $\Gamma_{1,1}$ are obtained so that these planes are normal to ω and w_0 or t_0. Points $N_{0,0}$, $N_{0,1}$ and $N_{1,0}$ have to be found to obtain $\Gamma_{1,1}$. $N_{0,0} = z \cap \omega$, $N_{0,1} \in t_0$, and $N_{0,1}$ is located it the distance d_{N01} = 5000 mm from $N_{0,0}$, and $N_{1,0} \in w_0$ as well as $N_{1,0}$ is lain in the distance d_{N10} = 4500 mm from $N_{0,0}$. The planes $\gamma_{w0,1}$ and $\gamma_{t1,0}$ pass through $N_{0,1}$ or $N_{1,0}$ perpendicularly to t_0 or to w_0. The unit vectors $n_{0,1} = [l_{n0,1}, m_{n0,1}, n_{n0,1}]$ and $n_{1,0} = [l_{n1,0}, m_{n1,0}, n_{n1,0}]$ are normal to ellipsoid ω. They determine, with the unit vectors $s_{t0,1}$ or $s_{w1,0}$ tangent to t_0 or w_0 at $N_{0,1}$ and $N_{1,0}$, two planes $\gamma_{w0,1}$ and $\gamma_{t1,0}$ of $\Gamma_{1,1}$. Thus, vectors $n_{1,1}$, $n_{0,1}$ and $n_{1,0}$ normal to ω at $N_{1,1}$, $N_{0,1}$ and $N_{1,0}$ are helpful in determining $\Gamma_{1,1}$. In order to determine vector $n_{0,1}$, vector $s_{w0,1} \parallel y$ can be passed through $N_{0,1}$, so $n_{0,1} = s_{w0,1} \times s_{t0,1}$ (Figure 14a). By analogy, $n_{1,0} = s_{w1,0} \times s_{t1,0}$. $N_{1,1}$ is created as a result of the intersection of ω with edge $h_{1,1} = \gamma_{w0,1} \cap \gamma_{t1,0}$. $\Gamma_{1,1}$ is symmetrical towards (x, z) or (y, z).

Side edge $h_{1,1}$ is not identical with straight line $n_{1,1}$ normal to ω at $N_{1,1}$ but only close to that line. An action leading to such a situation that the direction of $h_{1,1}$ is the closest possible to the direction of $n_{1,1}$ is expected. It may be obtained by changing the inclination of $\gamma_{w0,1}$ to (y, z) and the inclination of $\gamma_{t1,0}$ to (x, z), so the inclination of $h_{1,1}$ towards $n_{1,1}$ is also changed. The control of the above changes so that the angle of the inclination of $h_{1,1}$ to $n_{1,1}$ will be equal to the angles between the new and old positions of $\gamma_{w0,1}$ and $\gamma_{t1,0}$ is needed, however, this activity goes beyond the scope of this paper.

Reference tetrahedron $\Gamma_{1,1}$ is used for the central control composition of final reference structure Γ composed of nine reference tetrahedrons $\Gamma_{i,j}$. The plane $\gamma_{w0,1}$ of $\Gamma_{1,1}$ is accepted as one of four planes of the new reference tetrahedron $\Gamma_{1,2}$. The other three planes of $\Gamma_{1,2}$ are constructed in the following order (Figure 14b): (a) point $N_{0,2} \in t_0$ on ellipse t_0 determined in the distance $d_{w0,2} = 5000$ mm from $N_{0,1}$; (b) plane $\gamma_{w0,2}$ passing through $N_{0,2}$ and normal to t_0; (c) straight line $o_{w1,2} = \gamma_{w0,2} \cap \gamma_{w0,1}$; (d) straight line $o_{t1,2}$ passing through point $H_{t1,1} = h_{1,1} \cap h_{L1,1}$ and parallel to $(N_{0,1}, N_{0,2})$. On the basis of the above elements and activities, the following sets are obtained: a) the tetrad of planes: $\gamma_{w0,1}$, $\gamma_{w0,2}$, $\gamma_{t1,2}(o_{t1,2}$, $h_{1,1})$ and $\gamma_{Lt1,2}$ symmetrical to $\gamma_{t1,2}$ towards (x, z); and b) the tetrad of side edges of $\Gamma_{1,2}$: $h_{1,1}$, $h_{L1,1}$, $h_{1,2}$ = $\gamma_{t1,2} \cap \gamma_{w0,2}$ and $h_{L1,2}$ symmetrical to $h_{1,2}$ towards (x, z).

Reference tetrahedron $\Gamma_{2,1}$ is created in an analogous way as for $\Gamma_{1,2}$. Plane $\gamma_{t1,1}$ of $\Gamma_{1,1}$ is accepted as one of four planes of the new $\Gamma_{2,1}$ (Figure 15). The other three planes of $\Gamma_{2,1}$ are constructed by means of: a) point $N_{2,0} \in w_0$ constructed in the distance $d_{t2,0} = 4500$ mm from $N_{1,0}$; b) plane $\gamma_{t2,0}$ normal to w_0 and passing through $N_{2,0}$; c) straight line $o_{t2,1} = \gamma_{t1,0} \cap \gamma_{t2,0}$; and d) straight line $o_{w2,1}$ parallel to $(N_{1,0}, N_{2,0})$ and passing through point $H_{w1,1} = h_{1,1} \cap h_{p1,1}$. The following elements of $\Gamma_{2,1}$ are obtained: a) the tetrad of planes: $\gamma_{t1,0}$, $\gamma_{t2,0}$, $\gamma_{w2,1}$ $(o_{w2,1}$, $h_{1,1})$ and $\gamma_{pw2,1}$ symmetrical to $\gamma_{w2,1}$ towards (y, z); and b) the tetrad of side edges: $h_{1,1}$, $h_{p1,1}$, $h_{2,1} = \gamma_{w2,1} \cap \gamma_{t2,0}$ and $h_{p2,1}$ symmetrical to $h_{2,1}$ towards (y, z).

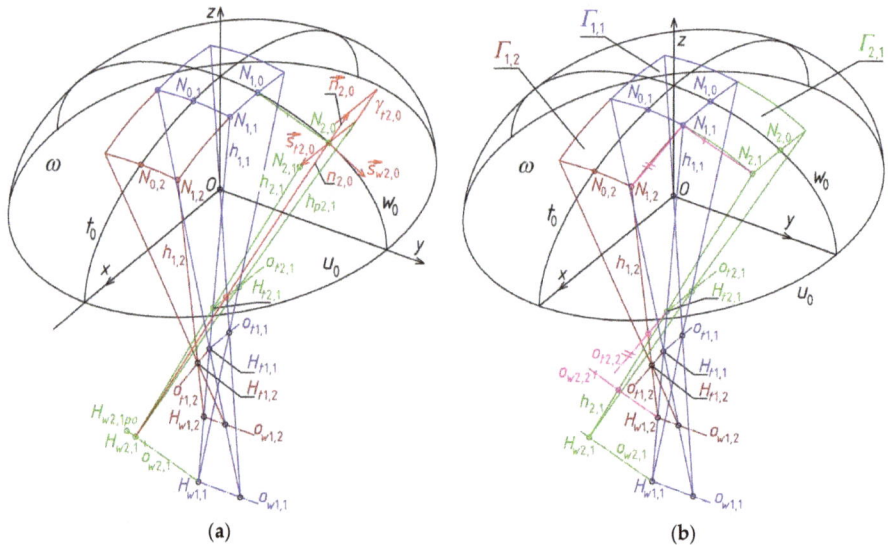

Figure 15. Constructions of a free form structure with the help of few reference tetrahedrons; (**a**) the auxiliary plane $\gamma_{t2,0}$ of $\Gamma_{2,1}$; (**b**) the reference tetrahedron $\Gamma_{2,1}$.

Reference tetrahedron $\Gamma_{2,2}$ is created in the way (Figure 16a) slightly different from the way used for $\Gamma_{1,2}$ and $\Gamma_{2,1}$ because two planes $\gamma_{w2,1}$ and $\gamma_{t1,2}$ from among the four planes of $\Gamma_{2,2}$ and three side edges $h_{1,1}$, $h_{1,2}$ and $h_{2,1}$ of $\Gamma_{2,2}$ have been obtained. In order to construct the fourth side edge $h_{2,2}$ of $\Gamma_{2,2}$ the following action should be executed. Straight line $o_{t2,2}$ is led through point $H_{t2,1} = h_{1,1} \cap h_{2,1}$ as parallel to straight line $(N_{1,1}, N_{1,2})$, where $N_{1,2} = h_{1,2} \cap w$. Straight line $o_{w2,2}$ is led through point $H_{w1,2} = h_{1,1} \cap h_{1,2}$ parallel to straight line $(N_{1,1}, N_{2,1})$, where $N_{2,1} = h_{2,1} \cap w$. Finally, points $H_{t2,2} = o_{t2,2} \cap h_{1,2}$ and $H_{w2,2} = o_{w2,2} \cap h_{2,1}$ determine edge $h_{2,2}$ (Figure 16b). Here, $h_{2,2}$ together with $h_{1,2}$ and $h_{2,1}$ determine two planes of $\Gamma_{2,2}$ being sought.

Four reference tetrahedrons $\Gamma_{i,j}$ (for $i, j = 1,2$) were constructed so far. The other five reference tetrahedrons of Γ (Figure 16b) can be obtained by transforming the above four tetrahedrons $\Gamma_{i,j}$ (for $i, j = 1,2$) so that (x, z) and (y, z) ary the symmetry planes of Γ.

The final roof shell structure Ω composed of nine shell sectors $\Omega_{i,j}$ contained in nine $\Gamma_{i,j}$ is constructed. The activities leading to the determination of shell structure Ω, being the sum of sectors $\Omega_{i,j}$ of ruled surfaces created on the basis of $\Gamma_{i,j}$ and $\Delta_{i,j}$ are similar to those ones presented earlier for single sector $\Omega_{1,1}$ contained in $\Gamma_{1,1}$. Ruled surfaces $\Omega_{i,j}$ are created on the basis of ω and positioned symmetrically towards (x, z) or (y, z) by analogy with the example described earlier. Roof structure Ω is shown in Figure 16b.

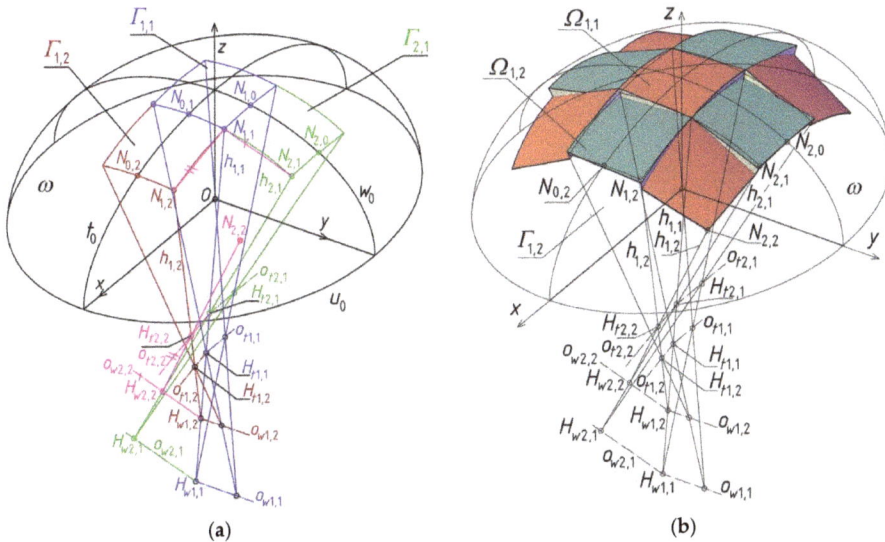

Figure 16. A free form structure created with the help of nine reference tetrahedrons: (**a**) Constructions of the reference tetrahedron $\Gamma_{2,2}$; (**b**) the simplified model of a shell roof structure.

Visualization of the achieved free form structure roofed with multi-segment shell structures is shown in Figure 17. It is possible to obtain many diversified, consistent architectural forms of such buildings from which two are presented in Figures 18 and 19. They are modifications of the structure created previously.

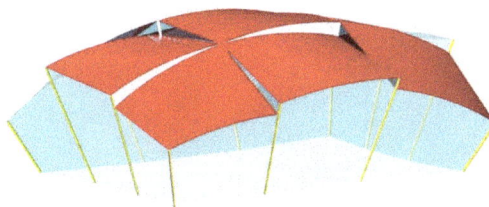

Figure 17. Visualization of the achieved free form structure roofed with a multi-segment shell structure.

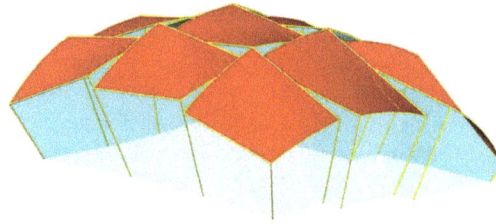

Figure 18. Visualization of the modified free form structure roofed with a multi-segment shell structure and folded elevation.

Figure 19. Visualization of another modified free form structure roofed with a multi-segment shell structure and folded elevation.

7. Optimized Structures Based on Regular Reference Surface

A non-rotational ellipsoid is used as a reference surface in the example presented below. The equation of this ellipsoid σ is the same as previously (1), but a = 24,000 mm, b = 18,000 mm, c = 11,000 mm. The gable wall of the designed structure can be located in one plane (Figure 20a) or divided into two planar pieces symmetrical towards the plane (x, z) (Figure 20b). The both forms are presented in the first part of the section. The shell roofs corresponding to these forms take very simple shapes. The roofs are created as the sums of a few shell strips whose directrices became the ellipses of the intersection of the gable wall plane or other almost vertical planes with the aforementioned reference ellipsoid σ.

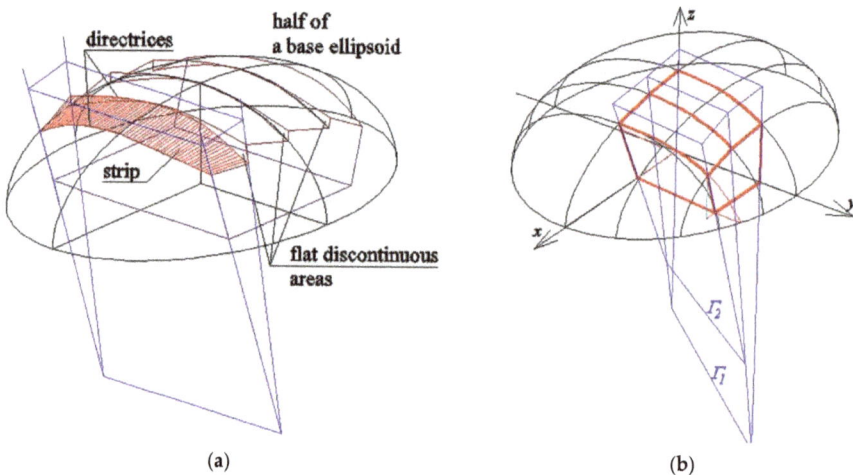

(a)   (b)

Figure 20. Geometric shaping of shell structures with a reference ellipsoid and reference tetrahedrons: (a) a single flat gable wall; (b) a symmetrical part of a gable wall.

The discussed method proposes to create more extended forms based on the reference surface σ, covered with compound shell structures supported by walls formed from many planar or shell segments, as shown in Figure 21. As a result, innovative, attractive and integrated building forms can be provided.

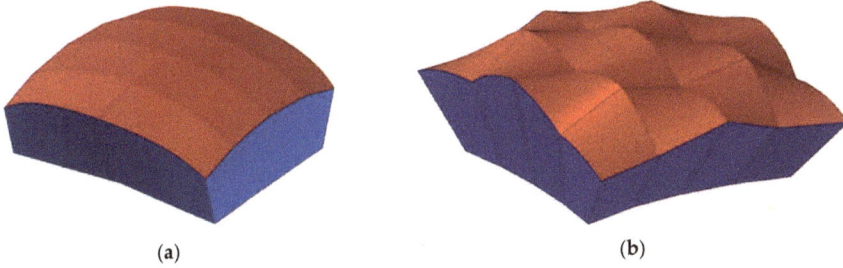

(a)   (b)

Figure 21. (**a**) Shell roof strips and the oblique gable wall composed of two parts contained in various planes; (**b**) Shell roof segments and the oblique gable wall composed of three parts contained in various planes.

If the number of the planar pieces of the gable wall is increased, then the integration of this wall with the entire building may be improved. In the next example, a regular reference structure is created. For that purpose, a finite number of points $N_{i,j}$ is defined on the reference ellipsoid with the help of ellipses w_i and t_i ($i = 0, 1, 2$) contained in vertical planes (Figure 22). The coordinates of the considered points $N_{i,j}$ ($i = 0, 1, 2, j = 0, 1, 2$) selected on the reference surface σ are included in Table 6.

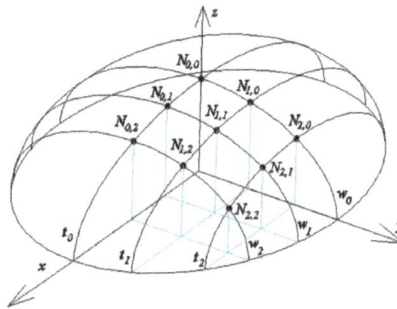

Figure 22. Visualization of the achieved free form structure roofed with a multi-segment shell structure.

Table 6. Coordinates of points $N_{i,j}$ selected on reference ellipsoid σ.

Vertex	X-Coordinate	Y-Coordinate	Z-Coordinate
$N_{0,0}$	0.0	0.0	11,000.0
$N_{1,0}$	0.0	6459.0	10,267.0
$N_{2,0}$	0.0	12,515.0	7906.0
$N_{0,1}$	6487.0	0.0	10,590.0
$N_{0,2}$	12,855.0	0.0	9289.0
$N_{1,1}$	6487.0	6459.0	9828.0
$N_{2,1}$	6487.0	12,515.0	7326.0
$N_{1,2}$	12,855.0	−6792.8	8409.0
$N_{2,2}$	12,855.0	6459.0	5272.0

The parametric equations of the considered ellipses t_i of reference ellipsoid σ are given by:

$$
\begin{aligned}
x &= a_t \cdot \cos(\tau_i) \\
y &= y_{Ni,0} \\
z &= c_t \cdot \sin(\tau_i),
\end{aligned}
\tag{2}
$$

where $a_t = a \cdot \sqrt{1 - \frac{y_{Ni,0}^2}{b^2}}$, $c_t = c \cdot \sqrt{1 - \frac{y_{Ni,0}^2}{b^2}}$, τ_i—the interdependent variable, $y_{Ni,0}$—y-coordinate of point $N_{i,0}$ ($i = 0, 1, 2$). The parametric equation of the ellipses w_j of σ are given by:

$$
\begin{aligned}
x &= x_{N0,j} \\
y &= b_w \cdot \cos(\omega_j) \\
z &= c_w \cdot \sin(\omega_j),
\end{aligned}
\tag{3}
$$

where $b_w = b \cdot \sqrt{1 - \frac{x_{N0,j}^2}{a^2}}$, $c_w = c \cdot \sqrt{1 - \frac{x_{N0,j}^2}{a^2}}$, ω_j - the interdependent variable, $x_{Ni,0}$ – x-coordinate of point $N_{0,j}$ ($j = 0, 1, 2$).

One plane of the reference network can be created for each pair of two subsequent lines $\{n_{i,j}, n_{i+1,j}\}$ normal to σ (Figure 23) to increase the integration degree of the entire structure with reference surface σ. On the basis of the above equations, straight lines $s_{tNi,j}$ and $s_{wNi,j}$ tangent to ellipses t_i and w_j of σ at $N_{i,j}$ are determined, (Figure 23a). Based on these tangents, the directional vectors of straight lines $n_{i,j}$ normal to this ellipsoid at $N_{i,j}$ were calculated. The values of the components $[l_{ni,j}, m_{ni,j}, m_{ni,j}]$ of these directional vectors are given in Table 7.

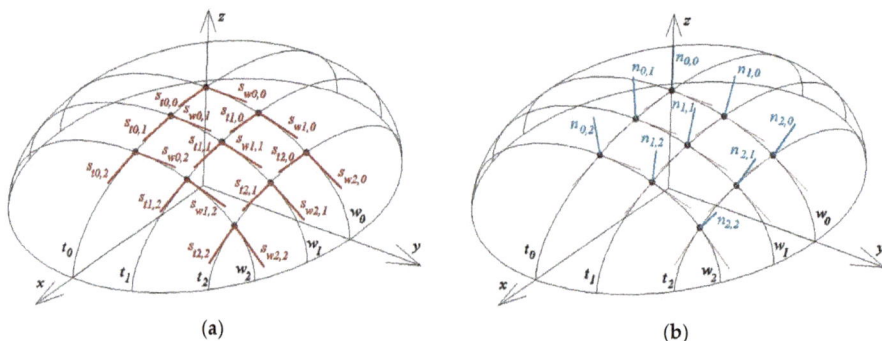

(a)   (b)

Figure 23. (a) Straight lines $s_{ti,j}$, $s_{wi,j}$ tangent to the reference ellipsoid at points spaced on its directrices t_i, w_j at constant distances; (b) Straight lines $n_{i,j}$ normal to the reference ellipsoid at the same points.

Table 7. Values of the components of the directional vectors of straight lines $n_{i,j}$ normal to the ellipsoid.

Vertex	$l_{ni,j}$	$m_{ni,j}$	$n_{ni,j}$
$N_{0,0}$	0.0	0.0	5000.0
$N_{1,0}$	0.0	1143.4	4867.5
$N_{2,0}$	0.0	2544.3	4304.2
$N_{0,1}$	638.1	0.0	4959.1
$N_{0,2}$	1395.8	0.0	4801.2
$N_{1,1}$	667.3	1181.0	4812.4
$N_{2,1}$	774.6	2656.6	4164.4
$N_{1,2}$	1474.8	1317.3	4592.3
$N_{2,2}$	1789.5	3097.1	3493.6

In order to construct the first tetrahedron $\Gamma_{1,1}$ of reference network Γ, whose side edges $H_{i-1,1}$ ($i = 1$–4) pass through points $N_{0,0}$, $N_{1,0}$, $N_{0,1}$, $N_{1,1}$, four pairs of straight lines $\{n_{0,0}, n_{0,1}\}$, $\{n_{0,1}, n_{1,1}\}$, $\{n_{1,0}, n_{1,1}\}$ and $\{n_{0,0}, n_{1,0}\}$ normal to σ at these points must be considered. Next, for each pair of these normals, the intersection of both lines with a straight line perpendicular to them should be constructed. For the pair $\{n_{0,1}, n_{1,1}\}$, there is a straight line $n_{Hw1,1}$ and points $H_{w0,1\_1,1} \in n_{0,1}$ and $H_{w,11\_1,1} \in n_{1,1}$. For the pair $\{n_{1,0}, n_{1,1}\}$ there is the straight line $n_{Ht1,1}$ and points $H_{t1,0\_1,1} \in n_{1,0}$ and $H_{t1,1\_1,1} \in n_{1,1}$ (Figure 24).

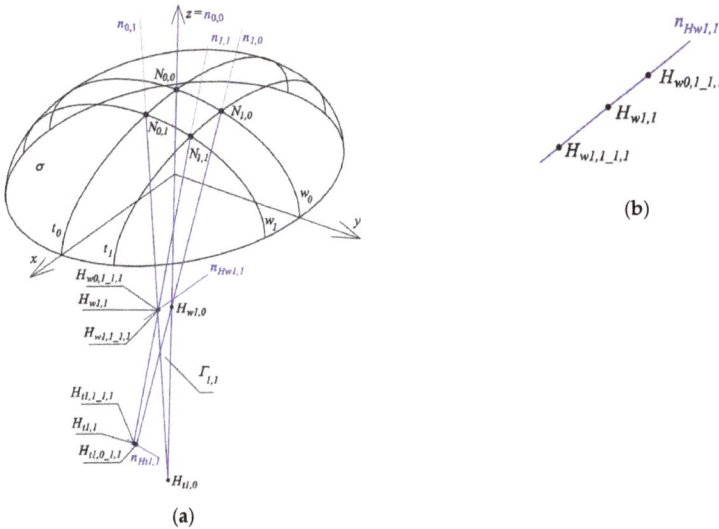

Figure 24. (a) Creation of the first reference tetrahedron $\Gamma_{1,1}$: (a) side edges and vertices, (b) search for one vertex $H_{w1,1}$.

For the other two cases: $\{n_{0,0}, n_{0,1}\}$, $\{n_{0,0}, n_{1,0}\}$, both straight lines of each of these pairs are coplanar, and therefore these lines cross each other at vertices $H_{w0,0}$ and $H_{t0,0}$ of the reference tetrahedron $\Gamma_{1,1}$, and it is not necessary to carry out relevant constructions. The notation of point $H_{w0,1\_1,1}$ should be interpreted as follows: (a) symbol w indicates that this element is related to curve w_j; (b) subscript 0,1 indicates that this point is referred to $N_{0,1}$; and (c) subscript 1,1 means that the considered point is used for determining $\Gamma_{1,1}$.

A tetrad of planes $\zeta_{tN0,1}$, $\zeta_{wN1,0}$, $\zeta_{tN1,1}$, $\zeta_{wN1,1}$ of tetrahedron $\Gamma_{1,1}$ is formed from two planes (x, z), (y, z) of the orthogonal coordinate system $[x,y,z]$ and two planes defined by the following triads of points: $(N_{1,0}, N_{1,1}, H_{t1,1})$, $(N_{0,1}, N_{1,1}, H_{w1,1})$, where $H_{t1,1}$ is the middle point of section $H_{t1,0\_1,1}H_{t1,1\_1,1}$, however, $H_{w1,1}$ is the middle point of section $H_{w0,1\_1,1}H_{w1,1\_1,1}$. Ultimately, edges $h_{0,0\_1,1} = \zeta_{tN0,1} \cap \zeta_{wN1,0}$, $h_{0,1\_1,1} = \zeta_{tN0,1} \cap \zeta_{wN1,1}$, $h_{1,0\_1,1} = \zeta_{tN1,1} \cap \zeta_{wN1,0}$, $h_{1,1\_1,1} = \zeta_{tN1,1} \cap \zeta_{wN1,1}$ are constructed. They are preliminary approximations of the side edges of the polyhedral reference structure Γ, and take positions close to the positions of straight lines $n_{i,j}$ normal to the arbitrary reference ellipsoid σ at points $N_{i,j}$ ($i = 0,1$, $j = 0,1$).

Each two subsequent straight lines from $\{h_{0,0\_1,1}, h_{0,1\_1,1}, h_{1,0\_1,1}, h_{1,1\_1,1}\}$ passing through adjacent points $N_{i,j}$ of ellipsoid σ intersect at four corresponding vertices $\{H_{w1,0}, H_{t1,0}, H_{w1,1}, H_{t1,1}\}$ of the tetrahedron $\Gamma_{1,1}$. Points $N_{i,j}$ ($i = 0,1$, $j = 0,1$) together with vertices $N_{i,j}$ define the four planes $\zeta_{tN0,1}$, $\zeta_{wN1,0}$, $\zeta_{wN1,1}$ and $\zeta_{tN1,1}$ of $\Gamma_{1,1}$.

In order to construct the second tetrahedron $\Gamma_{2,1}$, whose side edges pass through points $N_{1,0}$, $N_{1,1}$, $N_{2,1}$, $N_{2,0}$ (Figure 25), it is necessary to take into account the previously considered pair $\{n_{1,0}, n_{1,1}\}$ and next three pairs of straight lines $\{n_{2,0}, n_{2,1}\}$, $\{n_{1,1}, n_{2,1}\}$ and $\{n_{1,0}, n_{2,0}\}$.

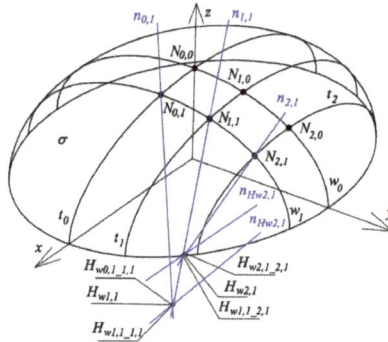

Figure 25. Creation of the second reference tetrahedron $\Gamma_{2,1}$.

Next, for each of these pairs, the points of both straight lines intersecting with a straight line perpendicular to them should be constructed. In the case of pair $\{n_{1,1}, n_{2,1}\}$, these are points $H_{w1,1\_2,1} \in n_{1,1}$ and $H_{w,21\_2,1} \in n_{2,1}$ and straight line $n_{Hw2,1}$ (Figure 25). In the case of pair $\{n_{2,0}, n_{2,1}\}$, these are points $H_{t2,0\_2,1} \in n_{2,0}$ and $H_{t2,1\_2,1} \in n_{2,1}$. Straight lines $\{n_{1,0}, n_{2,0}\}$ are coplanar and intersect at vertex $H_{w2,0}$ of tetrahedron $\Gamma_{2,1}$. The side facets of tetrahedron $\Gamma_{2,1}$ are: $\zeta_{tN1,1}, \zeta_{wN2,0}, \zeta_{tN2,1}, \zeta_{wN2,1}$, where $\zeta_{wN2,0}$ is the plane (y, z) of the coordinate system $[x, y, z]$, $\zeta_{tN1,1}$ was described earlier.

In turn, two planes: $\zeta_{tN2,1}, \zeta_{wN2,1}$ are defined respectively by the following triads of points: $(N_{2,0}, N_{2,1}, H_{t2,1})$, $(N_{1,1}, N_{2,1}, H_{w2,1})$, where $H_{t2,1}$ is the middle of section $H_{t2,0\_2,1}H_{t2,1\_1,1}$, and $H_{w2,1}$ is the middle of segment $H_{w1,1\_2,1}H_{w2,1\_2,1}$. Straight lines $h_{1,0\_2,1} = \zeta_{tN1,1} \cap \zeta_{wN2,0}$, $h_{1,1\_2,1} = \zeta_{tN1,1} \cap \zeta_{wN2,1}$, $h_{2,0\_2,1} = \zeta_{tN2,1} \cap \zeta_{wN2,0}$, $h_{2,1\_1,1} = \zeta_{tN2,1} \cap \zeta_{wN2,1}$ are initial approximations of straight lines $n_{Ni,j}$ normal to the reference surface σ at points $N_{i,j}$ ($i = 1, 2, j = 0, 1$) and are taken as preliminary approximations of the four side edges of tetrahedron $\Gamma_{2,1}$. Each two subsequent straight lines from the four following lines $\{h_{1,0\_2,1}, h_{1,1\_2,1}, h_{2,0\_2,1}, h_{2,1\_1,1}\}$ pass through adjacent points $N_{i,j}$ of ellipsoid σ and intersect in the appropriate four vertices $\{H_{w2,0}, H_{t2,0}, H_{w2,1}, H_{t2,1}\}$ of tetrahedron $\Gamma_{2,1}$ (Figure 26).

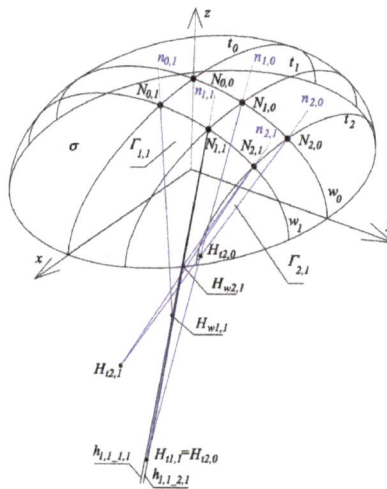

Figure 26. Creation of the sum of two adjacent tetrahedrons $\Gamma_{1,1}, \Gamma_{2,1}$ does not give a reference structure.

The sum of constructed tetrahedrons $\Gamma_{1,1}$ and $\Gamma_{2,1}$ does not create reference structure Γ, because, for example, straight lines $h_{1,1\_1,1}, \subset \Gamma_{1,1}$, i $h_{1,1\_2,1} \subset \Gamma_{2,1}$ are not identical (Figure 26), so the above tetrahedrons do not have a common side edge passing through point $N_{1,1}$. In this case, the set of three adjacent planes $\zeta_{tN1,1} \circ \zeta_{wN1,1} \circ \zeta_{wN2,1}$ does not have a common edge, so the above tetrahedrons do not share any common wall.

Therefore, some operation is necessary to replace the two above various straight lines with one straight line in an effective way, so that the above system was replaced by a new system of three planes $\zeta_{tN1,1n} \circ \zeta_{wN1,1n} \circ \zeta_{wN2,1n}$ having one common edge $k_{1,1}$. However, the position of this edge of the new system should be the closest possible to the position of normal $n_{1,1}$ of ellipsoid σ at point $N_{1,1}$. Therefore, an optimization process is necessary, so that sum S_{Min} of square of angles $\varphi_{i,j}$ between each plane $\zeta_{tNi,jn}$ or $\zeta_{wi,jn}$ of the new system and the corresponding plane $\zeta_{tNi,jn}$ or $\zeta_{wi,jn}$ of the old system was the smallest possible. The optimization condition reads:

$$\sum_{i=1}^{i_n=2} \sum_{j=1}^{j_n=2} \varphi_{Hri,jn}^2 \, S_{Min} = min, \tag{4}$$

where: $r = w$ or t, so it is obtained $\varphi_{i,jn}$ for points $H_{wi,jn}$ and $\zeta_{wi,jn}$, and $\varphi_{i,jn}$ for points $H_{ti,jn}$ and $\zeta_{tNi,jn}$.

It is then assumed that edge $k_{1,1}$ of the three new planes will be the closest possible to the line $n_{1,1}$ normal to ellipsoid σ within acceptable modeling accuracy. The description of the way of determining straight line $k_{1,1}$ will be presented after considering two next reference tetrahedrons $\Gamma_{1,2}, \Gamma_{2,2}$, because this straight line will replace four corresponding side edges of tetrahedrons $\Gamma_{1,1}, \Gamma_{2,1}, \Gamma_{1,2}, \Gamma_{2,2}$ with no common side edge passing through point $N_{1,1}$. The results of the optimization process performed for the reference structure Γ_n composed of four reference tetrahedrons $\Gamma_{i,jn}$ replacing $\Gamma_{i,j}$ ($i = 1.2, j = 1.2$) are also presented at the end of this section.

The last two considered tetrahedrons $\Gamma_{2,1}, \Gamma_{2,2}$ are defined in a manner analogous to the one presented earlier for $\Gamma_{1,1}$. Therefore, the following pairs of straight lines are investigated: $\{n_{0,2}, n_{1,2}\}$, $\{n_{1,1}, n_{1,2}\}$, $\{n_{1,2}, n_{2,2}\}$, $\{n_{2,1}, n_{2,2}\}$ and their intersecting points with a corresponding straight line perpendicular to each pair are sought. In the case of pair $\{n_{0,2}, n_{1,2}\}$, these are points $H_{w0,2\_1,2} \in n_{0,2}$ and $H_{w1,2\_1,2} \in n_{1,2}$. In the case of pair $\{n_{1,1}, n_{1,2}\}$, these are points $H_{t1,1\_1,2} \in n_{1,1}$ i $H_{t1,2\_1,2} \in n_{1,2}$ (Figure 27a). In the case of pair $\{n_{1,2}, n_{2,2}\}$, these are points $H_{w1,2\_2,2} \in n_{1,2}$ i $H_{w2,2\_1,2} \in n_{2,2}$. For pair $\{n_{2,1}, n_{2,2}\}$, these are points $H_{t2,1\_2,2} \in n_{2,1}$ i $H_{t2,2\_2,2} \in n_{2,2}$.

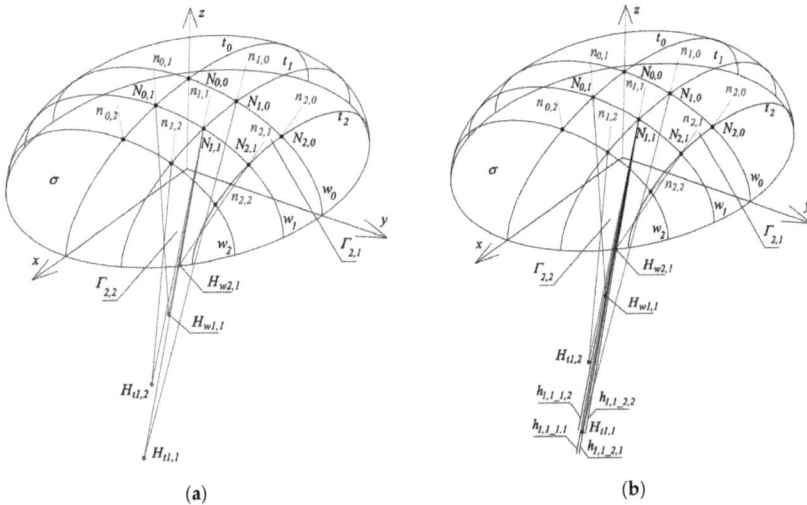

(a)　　　　　　　　　(b)

Figure 27. Construction of side edges of four reference tetrahedrons $\Gamma_{1,1}, \Gamma_{2,1}, \Gamma_{1,2}, \Gamma_{2,2}$: (**a**) various side edges; (**b**) search for one common side edge.

Tetrahedron $\Gamma_{1,2}$ is created by means of planes $\zeta_{tN0,2}$, $\zeta_{tN1,2}$, $\zeta_{wN1,2}$, $\zeta_{wN1,1}$, where $\zeta_{tN0,2}$ is the plane (x, z) of the coordinate system, $\zeta_{wN1,1}$ was defined before, while two planes $\zeta_{tN1,2}$, $\zeta_{wN1,2}$ are defined by the following triads of points: $(N_{1,1}, N_{1,2}, H_{t1,2})$, $(N_{0,2}, N_{1,2}, H_{w1,2})$, where $H_{t1,2}$ is the middle of $H_{t1,1\_1,2}H_{t1,2\_1,2}$ section, $H_{w1,2}$ is the middle of $H_{w0,2-\_1,2}H_{w1,2-\_1,2}$. Straight lines $h_{0,1\_1,2} = \zeta_{tN0,2} \cap \zeta_{wN1,1}$, $h_{0,2\_1,2} = \zeta_{tN0,2} \cap \zeta_{wN1,2}$, $h_{1,1\_1,2} = \zeta_{tN1,2} \cap \zeta_{wN1,1}$, $h_{1,2\_1,2} = \zeta_{tN1,2} \cap \zeta_{wN1,2}$ are adopted as the edges of tetrahedron $\Gamma_{1,2}$. These edges take positions close to the positions of straight lines $n_{i,j}$ normal to reference surface σ at points $N_{i,j}$ (for $i = 0,1, j = 1,2$). However, they are not side edges of the searched polyhedral reference structure because $\Gamma_{1,2}$ has no common edge with $\Gamma_{1,1}$ or $\Gamma_{2,1}$.

Tetrahedron $\Gamma_{2,2}$ is defined by means of planes $\zeta_{tN1,2}$, $\zeta_{tN2,2}$, $\zeta_{wN2,1}$, $\zeta_{wN2,2}$, where $\zeta_{tN1,2}$ and $\zeta_{wN2,1}$ were defined before. The next two planes $\zeta_{tN2,2}$, $\zeta_{wN2,2}$ should be determined using the following triplets of points: $(N_{2,1}, N_{2,2}, H_{t2,2})$, $(N_{1,2}, N_{2,2}, H_{w2,2})$, where $H_{t2,2}$ is the middle of segment $H_{t2,1\_2,2}H_{t2,2\_2,2}$, while H_{w2} is the middle of segment $H_{w1,2-\_2,2}H_{w2,2-\_2,2}$. The following straight lines $h_{1,1\_2,2} = \zeta_{tN1,2} \cap \zeta_{wN2,1}$, $h_{1,2\_2,2} = \zeta_{tN1,2} \cap \zeta_{wN2,2}$, $h_{2,1\_2,2} = \zeta_{tN2,2} \cap \zeta_{wN2,1}$, $h_{2,2\_2,2} = \zeta_{tN2,2} \cap \zeta_{wN2,2}$ are adopted as the edges of reference of tetrahedron $\Gamma_{2,2}$ and are the initial approximation of the edges of the polyhedral reference structure Γ.

The structure created as a result of adding the reference tetrahedrons $\Gamma_{1,1}$, $\Gamma_{2,1}$, $\Gamma_{1,2}$, $\Gamma_{2,2}$ is not reference structure Γ because, similarly as for the case of the sum of tetrahedrons $\Gamma_{1,1}$, $\Gamma_{2,1}$ considered earlier, two triads of planes: $\zeta_{tN1,2} \circ \zeta_{wN1,2} \circ \zeta_{wN2,2}$, $\zeta_{tN2,1} \circ \zeta_{wN2,1} \circ \zeta_{tN2,2}$ and four planes $\zeta_{tN1,1} \circ \zeta_{wN1,1} \circ \zeta_{wN2,1} \circ \zeta_{tN1,2}$ do not have common side edges. Thus, tetrahedrons $\Gamma_{1,2}$, $\Gamma_{2,2}$ and tetrahedrons $\Gamma_{2,1}$, $\Gamma_{2,2}$ as well as $\Gamma_{1,1}$, $\Gamma_{2,1}$, $\Gamma_{1,2}$, $\Gamma_{2,2}$ do not have appropriate common side edges passing through point $N_{2,1}$ or $N_{1,2}$ or $N_{1,1}$ (Figure 27b). Therefore, it is necessary to replace the above two triads and one tetrad with two new triads $\zeta_{tN1,2n} \circ \zeta_{wN1,2n} \circ \zeta_{wN2,2n}$, $\zeta_{tN2,1n} \circ \zeta_{wN2,1n} \circ \zeta_{tN2,2n}$ and one new tetrad $\zeta_{tN1,1n} \circ \zeta_{wN1,1n} \circ \zeta_{wN2,1n} \circ \zeta_{tN1,2n}$ in such a way that the position of the new planes and their edges is close to the positions of the respective straight lines $n_{i,j}$ normal to reference surface σ with the most possible precision.

Therefore, a second step of the initiated process of replacing straight lines $h_{r,s\_i,j}$ $(i, j, r, t = 1, 2)$ of planes $\zeta_{tNi,j}$ or $\zeta_{wNi,j}$ belonging to the created tetrahedrons $\Gamma_{i,j}$ not producing the polyhedral reference structure Γ, with side edges $k_{i,j}$ of planes $\zeta_{tNi,jn}$ and $\zeta_{wNi,jn}$ of the searched polyhedral reference structure Γ_n of several $\Gamma_{i,jn}$, is necessary. Since the above straight lines and planes should be as close as possible to straight lines $n_{i,j}$ normal to reference surface σ, it was assumed that sum S_{Min} (4) of squares of angles between the planes of the achieved new systems of planes $\zeta_{tN1,2n} \circ \zeta_{wN1,2n} \circ \zeta_{wN2,2n}$, $\zeta_{tN2,1n} \circ \zeta_{wN2,1n} \circ \zeta_{tN2,2n}$, $\zeta_{tN1,1n} \circ \zeta_{wN1,1n} \circ \zeta_{wN2,1n} \circ \zeta_{tN1,2n}$ and the corresponding planes of the previously created old systems of planes $\zeta_{tN1,2} \circ \zeta_{wN1,2} \circ \zeta_{wN2,2}$, $\zeta_{tN2,1} \circ \zeta_{wN2,1} \circ \zeta_{tN2,2}$, $\zeta_{tN1,1} \circ \zeta_{wN1,1} \circ \zeta_{wN2,1} \circ \zeta_{tN1,2}$ should be the smallest possible.

To calculate the angles between the $\zeta_{tNi,j}$, $\zeta_{wNi,j}$ planes determined in the first step for the tetrahedrons $\Gamma_{i,j}$ and $\zeta_{tNi,jn}$, $\zeta_{wNi,jn}$ for the $\Gamma_{i,jn}$ reference tetrahedrons forming the meshes of the searched reference network Γ_n, and estimated in the second step of the algorithm of the presented method the following formula was used

$$\varphi_{Hri,jn} = \frac{\pi}{2} - \text{asin}\left(\frac{n_{ki,j} \cdot n_{ki,jn}}{\left| n_{ki,j} \right| \cdot \left| n_{ki,jn} \right|}\right). \tag{5}$$

where $n_{ki,j}$ is the unit vector normal to plane $\zeta_{tNi,j}$ or $\zeta_{wNi,j}$, $n_{ki,jn}$ is the unit directional vector of $\zeta_{tNi,jn}$ or $\zeta_{wNi,jn}$, $\Pi = 3.14159$.

As a result of the optimization process, the following objects were obtained (Figure 28): (a) edge $k_{1,2}$ of three new planes $\zeta_{tN1,2n} \circ \zeta_{wN1,2n} \circ \zeta_{wN2,2n}$, which replaces two straight lines $h_{1,2\_1,2} \subset \Gamma_{1,2}$, $h_{1,2\_2,2} \subset \Gamma_{1,2}$ as accurately as possible; (b) edge $k_{2,1}$ of three new planes $\zeta_{tN2,1n} \circ \zeta_{wN2,1n} \circ \zeta_{tN2,2n}$, which replaces two straight lines $h_{2,1\_2,1} \subset \Gamma_{2,1}$, $h_{2,1\_2,2} \subset \Gamma_{2,2}$ as precisely as possible; and (c) edge $k_{1,1}$ of four new planes $\zeta_{tN1,1n} \circ \zeta_{wN1,1n} \circ \zeta_{wN2,1n} \circ \zeta_{tN1,2n}$, which substitutes four straight lines $h_{1,1\_1,1} \subset \Gamma_{1,1}$, $h_{1,1\_1,2} \subset \Gamma_{1,2}$, $h_{1,1\_2,1} \subset \Gamma_{2,1}$, $h_{1,1\_2,2} \subset \Gamma_{2,2}$ as exactly as possible. On the basis of the above planes, it is possible to determine edge $k_{2,2}$ of Γ_n as the straight line being the intersection of planes $\zeta_{wN2,2n}, \cap \zeta_{tN2,2n}$, where $\zeta_{tN2,2n}$ is defined by edge $k_{2,1}$ and point $N_{2,2}$, and $\zeta_{wN2,2n}$ is defined by edge $k_{1,2}$ and point $N_{2,2}$. As a result, the straight lines $k_{1,1}, k_{1,2}, k_{2,1}, k_{2,2}$, whose positions are very close to the positions of $n_{1,1}, n_{1,2}, n_{2,1}, n_{2,2}$ normal to ellipsoid σ, were obtained.

During the process of optimizing the positions of the facets and edges of reference tetrahedrons $\Gamma_{i,jn}$ to the positions of straight lines $n_{i,j}$ normal to the reference surface σ, the positions of the following points were changed: $H_{w1,1}, H_{w2,1}, H_{w2,2}, H_{t2,1}$ respectively on lines $(H_{w0,1\_1,1}, H_{w1,1\_1,1})$, $(H_{w1,1\_2,1}, H_{w2,1\_2,1})$, $(H_{w1,2\_\_2,2}, H_{w2,2\_2,2})$, $(H_{t2,0\_2,1}, H_{t2,1\_2,1})$. As a result, their new positions $H_{w1,1n}$, $H_{w2,1n}, H_{w2,2n}, H_{t2,1n}$ made it possible to build triples of points $(N_{0,1}, N_{1,1}, H_{w1,1n})$, $(N_{1,1}, N_{2,1}, H_{w2,1n})$, $(N_{1,2}, N_{2,2}, H_{w2,2n})$, $(N_{2,0}, N_{2,1}, H_{t2,1n})$ determining the planes $\zeta_{wN1,1n}, \zeta_{wN2,1n}, \zeta_{wN2,2n}, \zeta_{tN2,1n}$ sought. The location of points $H_{w1,1}, H_{w2,1}, H_{w2,2}, H_{t2,1}$ was controlled by parameters $w_{Hw1,1n}$, $w_{Hw2,1n}, w_{Hw2,2n}, w_{Ht2,1n}$, that is the division coefficients of sections $H_{w0,1\_1,1}H_{w1,1\_1,1}$, $H_{w1,1\_2,1}H_{w2,1\_2,1}$, $H_{w1,2\_\_2,2}H_{w2,2\_2,2}$ i $H_{t2,0\_2,1}H_{t2,1\_2,1}$, in contrast to the vertices $H_{w1,1}, H_{w2,1}, H_{w2,2}, H_{t2,1}$ adopted in the middles of the respective sections considered in the previous step. For other points $H_{wi,jn}, H_{ti,jn}$, the values of the division coefficients are the result of the optimization process and depend on the aforementioned four coefficients.

The algorithm of defining the optimal reference structure Γ_n based on the obtained points $H_{w1,1}$, $H_{w2,1}, H_{w2,2}, H_{t2,1}$ is presented below (Figure 28). Structure Γ_n is the sum of reference tetrahedrons $\Gamma_{i,jn}$, whose edges $k_{i,j}$ and planes $\zeta_{tNi,jn}$ can be achieved on the basis of the above four optimized division coefficients in the following way.

Points $N_{1,1}$ and $H_{w1,1n}$ define edge $k_{1,1}$. Straight line $k_{1,1}$ and point $N_{0,1}$ determine plane $\zeta_{wN1,1n}$ of the reference structure Γ_n. Analogously, the following planes: $\zeta_{wN2,1n}, \zeta_{tN1,1n}, \zeta_{tN1,2n}$ of structure Γ_n pass through edge $k_{1,1}$ and points $N_{1,0}, N_{1,2}, N_{2,1}$, respectively. Similarly, plane $\zeta_{wN2,2n}$ passing through points $N_{1,2}, N_{2,2}, H_{w2,2n}$ is the plane of structure Γ_n. Edge $k_{1,2}$ of structure Γ_n is the intersection of planes $\zeta_{tN1,2n}, \zeta_{wN2,2n}$. Edge $k_{1,2}$ and point $N_{0,2}$ define plane $\zeta_{wN1,2n}$. Planes $\zeta_{tN2,1n}$ and $\zeta_{wN2,1n}$ intersect each other in edge $k_{2,1}$. Plane $\zeta_{tN2,2n}$ is determined by points $N_{2,0}, N_{2,1}$ and $H_{t2,1n}$. Plane $\zeta_{wN2,2n}$ is determined by points $N_{1,2}, N_{2,2}, H_{w2,2n}$. Edge $k_{2,2}$ is the intersection of planes $\zeta_{tN2,2n}$ and $\zeta_{wN2,2n}$.

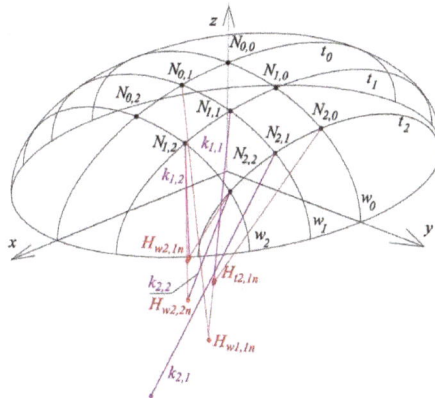

Figure 28. Creation of the side edges of reference structure Γ_n composed of four reference tetrahedrons $\Gamma_{1,1n}, \Gamma_{2,1n}, \Gamma_{1,2n}, \Gamma_{2,2n}$.

Edges $k_{0,1}$ and $k_{0,2}$ of structure Γ_n are the straight lines of the intersection of planes $\zeta_{wN1,1n}, \zeta_{wN1,2n}$ with plane (x, z). Edges $k_{1,0}$ and $k_{2,0}$ of Γ_n are the straight lines of the intersection of planes $\zeta_{tN1,1n}$, $\zeta_{tN2,1n}$ with plane (y, z). Axis z is adopted as edge $k_{0,0}$ of Γ_n. The reference tetrahedron $\Gamma_{2,2}$ (Figure 29) is created as the last part of the one-fourth of Γ_n.

The reference structure Γ_n is the sum of all reference tetrahedrons $\Gamma_{i,jn}$ whose walls, contained in the aforementioned planes $\zeta_{wNi,jn}$, $\zeta_{tNi,jn}$ ($i = 0, 1, 2$ and $j = 0, 1, 2$), are common to each pair of the adjacent reference tetrahedrons, and side edges $k_{i,j}$ are the shared corners of pairs, triples or tetrads of the neighboring reference tetrahedrons. The results of the optimization process performed for one of the four quarters of the considered reference structure Γ_n symmetrical towards two planes of coordinate system [x, y, z] are presented in Figure 30.

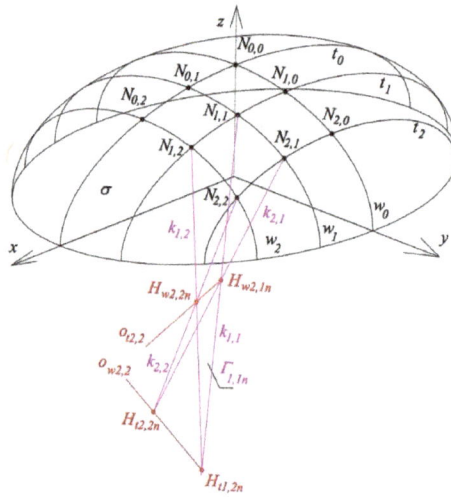

Figure 29. Creation of reference tetrahedron $\Gamma_{2,2n}$ of Γ_n structure.

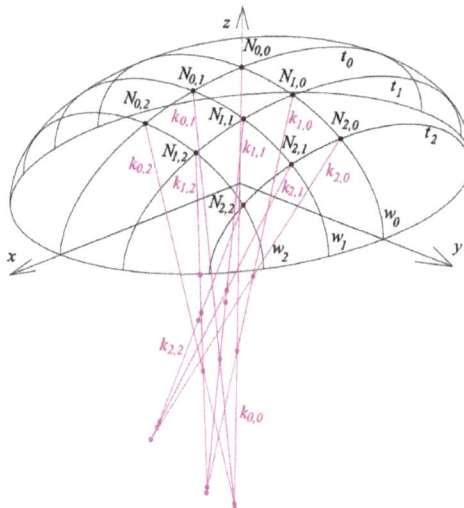

Figure 30. One-fourth of reference structure Γ_n, composed of four reference tetrahedrons $\Gamma_{1,1n}$, $\Gamma_{2,1n}$, $\Gamma_{1,2n}$, $\Gamma_{2,2n}$.

The values of selected coefficients $w_{Hwi,jn}$ or $w_{Hti,jn}$ evaluated in the presented iterative optimization process are included in Table 8. The obtained values of some angles between planes $\zeta_{tNi,j}$,

$\zeta_{wNi,j}$ of old tetrahedrons $\Gamma_{i,j}$ and planes $\zeta_{tNi,jn}$, $\zeta_{wNi,j}$ of new reference tetrahedrons $\Gamma_{i,jn}$ are included in Table 9. The investigated one-fourth of the structure Γ_n is composed of four reference tetrahedrons $\Gamma_{i,jn}$ ($i = 1, 2$ and $j = 1, 2$). The values of components $l_{ki,j}$, $m_{ki,j}$, $m_{ki,j}$ of the directional vectors of side edges $k_{i,j}$ of structure Γ_n, passing through points $N_{i,j}$ are given in Table 10.

Table 8. Division coefficients subsequently accepted in the iterative optimization process.

Iteration Step [No]	$w_{Hw1,1n}$	$w_{Hw2,1n}$	$w_{Hw2,2n}$	$w_{Ht2,1n}$
1	0.50	0.50	0.50	0.50
2	1.00	0.60	0.50	0.50
3	2.00	0.70	0.50	0.50
4	3.00	0.80	0.50	0.50
5	4.55	0.95	0.50	0.50
6	4.55	0.95	0.20	0.50
7	4.58	0.95	0.21	0.50

Table 9. Decreasing values of sum S_{Min} of square of angles $\varphi_{i,j}$ between planes $\zeta_{tNi,j}$, $\zeta_{wNi,j}$ of the old tetrahedrons $\Gamma_{i,j}$ and planes $\zeta_{tNi,jn}$, $\zeta_{wNi,j}$ of the new reference tetrahedrons $\Gamma_{i,jn}$ of the described iterative optimization process.

Iteration Step [No]	$\varphi_{Hw1,1n}$ [°]	$\varphi_{Hw2,1n}$ [°]	$\varphi_{Hw2,2n}$ [°]	$\varphi_{Ht1,1n}$ [°]	$\varphi_{Ht1,2n}$ [°]	$\varphi_{Ht2,1n}$ [°]	S_{Min}
1	0.00	0.00	0.00	20.17	21.23	0.00	857.7
2	0.17	0.13	0.00	18.85	19.90	0.16	751.4
3	0.52	0.26	0.00	13.42	14.47	0.80	390.2
4	0.86	0.39	0.00	8.02	9.08	1.41	149.6
5	1.40	0.58	0.00	0.20	0.88	2.34	8.6
6	1.40	0.58	1.23	0.20	0.88	1.12	5.9
7	1.41	0.58	1.19	0.44	0.64	1.19	5.7

Table 10. Values of components $l_{ki,j}$, $m_{ki,j}$, $m_{ki,j}$ of the unit directional vectors of side edges $k_{i,j}$ of Γ_n.

Vertex	$l_{ki,j}$	$m_{ki,j}$	$n_{ki,j}$
$k_{0,0}$	0.000	0.000	1.000
$k_{1,0}$	0.000	0.240	0.971
$k_{0,1}$	0.155	0.000	0.988
$k_{1,1}$	0.155	0.240	0.959
$k_{2,0}$	0.000	0.549	0.836
$k_{2,1}$	0.153	0.549	0.822
$k_{0,2}$	0.307	0.000	0.952
$k_{1,2}$	0.308	0.239	0.921
$k_{2,2}$	0.312	0.551	0.774

The directrices of each shell segment of the searched roof structure based on ellipsoid σ are sums of sections $w_{i,j}$ of arbitrary ellipses, for instance one directrix corresponding to w_2 is the sum of $w_{1,2} = \zeta_{wN1,2n} \cap \sigma$ and $w_{2,2} = \zeta_{wN2,2n} \cap \sigma$ (Figure 31). In addition, $w_{1,0}$ $w_{2,0}$ w_0, $w_{1,1} \neq w_1$, $w_{2,1} \neq w_1$, $\{w_{1,1g}, w_{1,1}\}$ $\zeta_{wN1,1n}$, $\{w_{2,1g}, w_{2,1}\}$ $\zeta_{wN2,1n}$, $w_{1,2} \neq w_{2,2} \neq w_2$. The lines $w_{1,1}$ and $w_{1,1g}$ as well as $w_{2,1}$ and $w_{2,1g}$ are coplanar sections of the directrices. The index g denotes that the proper curve e.g. $w_{2,1g}$ is located outside of the reference ellipsoid. It is possible to obtain such a structure that the conditions $w_{1,1} = w_{1,1g} = \zeta_{wN1,1n} \cap \sigma$ and $w_{2,1} = w_{2,1g} = \zeta_{w2,1} \cap \sigma$ are met. The visualization of the resultant architectural form of the discussed structure is presented in Figure 32.

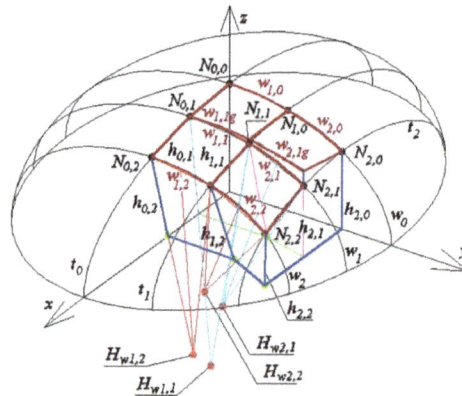

Figure 31. One of four parts of the reference structure and roof shell structure symmetrical towards (x, z) and (y, z) planes.

Figure 32. Visualization of the optimized free form structure.

8. Conclusions

Despite the relatively great possibilities of the search for diverse single free forms of buildings roofed with transformed shells, resulting from the freedom in selecting the shape and position of the roof directrices, there are significant limitations in creating these forms due to the geometrical and mechanical properties of the folded steel sheets. In order to overcome these limitations, the author proposed various methods for shaping the buildings as free form structures composed of many individual forms connected with common walls. Positive e effect of the skillful composition of many single warped surfaces for roofing is that the designed building free form structure is becomes internally consistent and externally sensitive to the built or natural environments. The possibility of further modification of these structures by means of displacements of roof directrices and elevation edges in the planes of the auxiliary reference tetrahedrons, defined by the author, allows the aforementioned internal coherence and external sensitivity to be increased.

Three methods of creating composite building free forms roofed with structures of many shell segments made up of transformed corrugated sheeting are proposed. Based on the results of studies on the first method, it can be concluded that the reference tetrahedrons and operations proposed in the algorithm of this method enable easy and creative creation of such complex free forms characterized by integrated forms of roofs and façades.

Moreover, it is very easy to modify these complete tetrahedrons in order to obtain many different configurations of the free forms sensitive to the natural and built environments. This modification consists of: (1) changing the position of the roof eaves' corners along the side edges of the façade walls in order to change the mutual position of shell roof segments; and (2) changing the position of the vertices of the reference tetrahedrons along the axes of the tetrahedrons to obtain the corrugation of flat façade walls.

The algorithm of the second method introduces a certain regularity in the placement and joining of subsequent reference tetrahedrons in the three-dimensional space into one regular spatial polyhedral reference network. To achieve this regularity, an auxiliary reference surface is introduced as a double-curved regular surface whose specific properties are used to build and arrange the reference tetrahedrons which are the meshes of the reference network. The algorithm is of no particular support for the designer because it does not offer additional conditions, allowing the form of the reference network to be regular and take into account the variable curvature of the reference surface.

Such additional conditions, effectively supporting the designer's activity, are provided by the very sophisticated third method proposed by the author. The method replaces straight lines normal to the reference surface with side edges of the searched reference network. However, each pair of the adjacent side edges of the reference network must intersect, while the respective two straight lines normal to the reference surface are skewed. Therefore, to solve this problem, the algorithm of the method is based on the optimization of the directions of several side edges of the reference network in relation to a finite number of selected straight lines normal to the reference surface. As a result, the differences in the directions of the side edges and corresponding normals are as small as possible.

The algorithm uses an optimization process, the idea of which is to search for the positions of selected planes of the reference network so that the position of each plane was the closest possible to two subsequent normals to the reference surface. The obtained pairs of subsequent planes have to intersect at the side edges of the reference network that approximate the position of the above normals to the reference surface.

Each plane of the reference network is defined by means of three points. Two of these are points of the intersection of two subsequent normals with the reference surface. The third point is sought on a straight line perpendicular to the above two normals, and intersecting these normals. Therefore, the position of this point is optimized on the aforementioned straight line, for several planes of the reference network. The result of the optimization carried out in the article indicates that the optimal position of each such a point is not, as might be expected, the middle of the section with its ends at the intersecting points of the above three straight lines, but this position is dependent on the changes in curvature on the reference surface and must be calculated during the optimization process.

Obviously, this location is determined by the variability of the curvatures of the reference surface. Therefore, in the future, the author intends to develop a parametric description of the relationship between the overall dimensions and curvatures of the arbitrary smooth regular reference surface and the properties of the optimized reference network searched for the reference surface. In addition, this description should take into account the choice of other characteristic lines on the reference surface, such as geodesic or curvature lines. This description will allow writing a relevant computer application supporting the designer in shaping complex building free forms.

Funding: The resources of the Rzeszow University of Technology.

Conflicts of Interest: The author declares no conflict of interest.

References

1. Foraboschi, P. The central role played by structural design in enabling the construction of buildings that advanced and revolutionized architecture. *Constr. Build. Mater.* **2016**, *114*, 956–976. [CrossRef]
2. Foraboschi, P. Structural layout that takes full advantage of the capabilities and opportunities afforded by two-way RC floors, coupled with theselection of the best technique, to avoid serviceability failures. *Eng. Fail. Anal.* **2016**, *70*, 377–418. [CrossRef]
3. Abel, J.F.; Mungan, I. *Fifty Years of Progress for Shell and Spatial Structures*; International Association for Shell and Spatial Structures: Madrid, Spain, 2011.
4. Foraboschi, P. Optimal design of glass plates loaded transversally. *Mater. Des.* **2014**, *62*, 443–458. [CrossRef]

5. Liu, Y.; Zwingmann, B. Carbon Fiber Reinforced Polymer for Cable Structures—A Review. *Polymers* **2015**, *7*, 2078–2099. [CrossRef]

6. Adriaenssens, S.; Ney, L.; Bodarwe, E.; Williams, C. Finding the form of an irregular meshed steel and glass shell based on construction constraints. *Archit. Eng.* **2012**, *18*, 206–213. [CrossRef]

7. Makowski, Z.S. *Analysis, Design and Construction of Double-Layer Grids*; Applied Science Publishers: London, UK, 1981.

8. Medwadowski, S.J. Symposium on Shell and Spatial Structures: The Development of Form. *Bull. IASS* **1979**, *70*, 3–10.

9. Saitoh, M. *Recent Spatial Structures in Japan*; J. JASS: Madrid, Spain, 2001.

10. Vizotto, Computational generation of free-form shells in architectural design and civil engineering. *Autom. Constr.* **2010**, *19*, 1087–1105. [CrossRef]

11. Reichhart, A. *Geometrical and Structural Shaping Building Shells Made up of Transformed Flat Folded Sheets*; Rzeszow University of Technology: Rzeszów, Poland, 2002. (In Polish)

12. Bletzinger, K.U.; Wüchner, R.; Kupzok, A. Algorithmic treatment of shells and free form-membranes in FSI. *Fluid-Struct. Interact.* **2006**, *11*, 336–355.

13. Abramczyk, J. *Shell Free forms of Buildings Roofed with Transformed Corrugated Sheeting*; Rzeszow University of Technology: Rzeszów, Poland, 2017.

14. Abramczyk, J. Shape transformations of folded sheets providing shell free forms for roofing. In Proceedings of the 11th Conference on Shell Structures Theory and Applications, Gdańsk, Poland, 11–13 October 2017; Pietraszkiewicz, W., Witkowski, W., Eds.; CRC Press Taylor and Francis Group: Boca Raton, FL, USA, 2017; pp. 409–412.

15. Bathe, K.J. *Finite Element Procedures*; Prentice Hall: Englewood Cliffs, NJ, USA, 1996.

16. Abramczyk, J. Shaping Innovative Forms of Buildings Roofed with Corrugated Hyperbolic Paraboloid Sheeting. *Procedia Eng.* **2016**, *8*, 60–66. [CrossRef]

17. Abramczyk, J. Integrated building forms covered with effectively transformed folded sheets. *Procedia Eng.* **2016**, *8*, 1545–1550. [CrossRef]

18. Prokopska, A.; Abramczyk, J. Responsive Parametric Building Free Forms Determined by Their Elastically Transformed Steel Shell Roofs Sheeting. *Buildings* **2019**, *9*, 46. [CrossRef]

19. Abramczyk, J. Building Structures Roofed with Multi-Segment Corrugated Hyperbolic Paraboloid Steel Shells. *J. Int. Assoc. Shell Spat. Struct.* **2016**, *2*, 121–132. [CrossRef]

20. Reichhart, A. Corrugated Deformed Steel Sheets as Material for Shells. In Proceedings of the I International Conference on Lightweight Structures in Civil Engineering, Warsaw, Poland, 26–29 December 1995.

21. Grey, A. *Modern Differential Geometry of Curves and Surfaces with Mathematica*, 4th ed.; Champman & Hall: New York, NY, USA, 2006.

22. Abramczyk, J. Principles of geometrical shaping effective shell structures forms. *JCEEA* **2014**, *XXXI*, 5–21. [CrossRef]

23. Viskovi, A. Mode Pressure Coefficient Maps as an Alternative to Mean Pressure Coefficient Maps for Non-Gaussian Processes: Hyperbolic Paraboloid Roofs as Cases of Study. *Computation* **2018**, *6*, 64. [CrossRef]

24. Gergely, P.; Banavalkar, P.V.; Parker, J.E. The analysis and behavior of thin-steel hyperbolic paraboloid shells. In *A Research Project Sponsored by the America Iron and Steel Institute*; Report 338; Cornell University: Ithaca, NY, USA, 1971.

25. Davis, J.M.; Bryan, E.R. *Manual of Stressed Skin Diaphragm Design*; Granada: London, UK, 1982.

26. Egger, H.; Fischer, M.; Resinger, F. *Hyperschale aus Profilblechen*; Der Stahlbau, H., Ed.; Ernst&Son: Berlin, Germany, 1971; Volume 12, pp. 353–361.

27. Makowski, Z.S. *Steel Space Structures*; Michael Joseph: London, UK, 1965.

28. Graham, L.; Scott, M.; Pappas, A. Natatorium Building Enclosure Deterioration Due to Moisture Migration. *Buildings* **2012**, *2*, 534–541. [CrossRef]

29. Sibley, M. Let There Be Light! Investigating Vernacular Daylighting in Moroccan Heritage Hammams for Rehabilitation, Benchmarking and Energy Saving. *Sustainability* **2018**, *10*, 3984. [CrossRef]

30. Gürlich, D.; Reber, A.; Biesinger, A.; Eicker, U. Daylight Performance of a Translucent Textile Membrane Roof with Thermal Insulation. *Buildings* **2018**, *8*, 118. [CrossRef]

31. Abramczyk, J. Parametric shaping of consistent architectural forms for buildings roofed with corrugated shell sheeting. *J. Archit. Civ. Eng. Environ.* **2017**, *10*, 5–18.

32. Prokopska, A.; Abramczyk, J. Parametric Creative Design of Building Free Forms Roofed with Transformed Shells Introducing Architect's and Civil Engineer's Responsible Artistic Concepts. *Buildings* **2019**, *9*, 58.

33. Prokopska, A.; Abramczyk, J. Innovative systems of corrugated shells rationalizing the design and erection processes for free building forms. *J. Archit. Civ. Eng. Environ.* **2017**, *10*, 29–40. [CrossRef]

34. Prokopska, A. *Methodology of Architectural design Preliminary Phases of the Architectural Process*; Publishing House of Rzeszow University of Technology: Rzeszów, Poland, 2018.

35. Prokopska, A. Creativity Method applied in Architectural Spatial cubic Form Case of the Ronchamp Chapel of Le Corbusier. *J. Transdiscipl. Syst. Sci.* **2007**, *12*, 49–57.

36. Obrębski, J.B. Observations on Rational Designing of Space Structures. In Proceedings of the Symposium Montpellier Shell and Spatial Structures for Models to Realization IASS, Montpellier, France, 20–24 September 2004; pp. 24–25.

37. Rębielak, J. Review of Some Structural Systems Developed Recently by help of Application of Numerical Models. In Proceedings of the XVIII International Conference on Lightweight Structures in Civil Engineering, Łódź, Poland, 7 December 2012; pp. 59–64.

buildings

MDPI

Article

Interaction Narratives for Responsive Architecture

Henri Achten

Department of architectural modelling, Faculty of architecture, Czech Technical University in Prague,
166 29 Prague, Czech Republic; achten@fa.cvut.cz

Received: 4 January 2019; Accepted: 12 March 2019; Published: 14 March 2019

Abstract: In this position paper, we present the results of an ongoing theoretical investigation into the phenomenon of interactive architecture. Interaction in architecture deals with the meaningful exchange of information and physical acts between building and person. This goes beyond responsive systems like automated doors, shading systems, and so on. Most examples of interactive architecture are technological explorations that probe possibilities and the potential for interaction. In this paper we claim that this is not enough. The notion of interactive architecture is explored through social aspects, user experience, situatedness, and agent-based theory. From this we argue that interactive buildings need comprehensive and consistent styles of interaction rather than a series of isolated and unrelated interaction events. Different people in various contexts require different sets of behavior from an interactive building. These sets are conceptualized as interaction narratives, following the work of Maria Lehman. We argue that such narratives can provide a better fit of the interactive building with the user, and lead to a more profound understanding of such systems.

Keywords: interactive architecture; interaction narrative; user experience; situatedness; agent-based theory

1. Introduction

Buildings rarely are conceived from the very first step of their design as things that must change a lot. This is a rather strange contradiction with the long time-span that buildings fulfill their function. During this lifetime changes occur along various time scales—the inhabitants will change, usually in cycles of decades; the needs of the inhabitants change in cycles of years; their activities and ways of using the building change per season; they have their weekly and daily rhythms, and during the day many different things happen inside the building. This applies to all types of buildings, domestic, work, industry, entertainment, and so on. Usually buildings are conceived as more or less unchanging containers in which these dynamics take place, without the need of changing the building in a drastic way. Changing buildings is costly, takes much time, and labor; therefore, physically changing buildings is avoided rather than embraced. Conventional design methods are ill-equipped to take changes described above into account, nor are there methods able to deliver building designs that appropriately incorporate such changes. Some attempts have been undertaken to deal with change, mainly through conventional means—like division of the building into parts with various time-spans or assigning a mandate for change to different parties who have a stake in the building [1].

In this paper, we are concerned with advanced technologies that make buildings much more dynamic than the traditional notion of buildings outlined above. Our perspective comes from the architect. We aim to develop a framework in which interaction is an integrated part when conceiving building designs. Thus, we are less concerned with the technical realization of interactive systems than the question, what are interactive buildings, and how are they designed. The notion of interactive buildings, which attributes much more dynamics to the building itself, has its roots in the work of Cedric Price (1935–2003) and Gordon Pask (1928–1996). The Interaction Centre, Kentish Town, by architect Cedric Price, realized in 1971, was one of the first contemporary examples of kinetic and flexible architecture. It was a more modest version of his earlier vision of the Fun Palace (together with

Joan Littlewood in 1961), an unrealized educational structure, which could be changed according to the desires of the visitors [2] (pp. 60). Already in 1969, Gordon Pask argued that architects design systems, rather than buildings, and should therefore consult cybernetics to design such buildings [3].

Advances in contemporary technology bring the notion of interactive architecture to a completely new level. It has become indeed possible to create building components that can move, change shape, or do other things that make buildings much more flexible and adaptable in easy ways. This means that the design of buildings shifts from the design of a more or less passive envelope and space program, to the question for the building, "what should I do next?" This question, and the answer to this question, is beyond contemporary architectural (design) theory. However, Pask's suggestion 50 years ago can now be reframed in the contemporary theory of agent-based systems [4], a concept that originates from artificial intelligence. According to Russell and Norvig [5], agents are components that can sense the world they are in, and affect change in that world, based on their autonomous reasoning. They define the so-called "ideal rational agent" as one that: "For each possible percept sequence, do whatever action is expected to maximize performance measure, on the basis of evidence provided by the percept sequence and built-in knowledge of the agent" (Ibid., page 23). This definition, however abstract, included important clues for the design of interactive buildings:

- Percept sequence: the sequence of events that is registered by the agent, in our case, the building. The implication is twofold; first, interactive buildings need to sense a relevant range of things happening in and around the building, and second, these series of events form a sequence that through a consistent interpretation can lead to meaningful inferences to what is happening.
- Action: anything that the agent may change in the world. The implication is that interactive buildings are not passive receivers of events, but actors by themselves.
- Maximize performance: the vector of quantifiable variables that sets the value of how well the agent is doing. For a rational agent, the choice what to do next is determined by which way a better value of the performance can be obtained. The extension of this concept to interactive buildings challenges the question what performance is. Since buildings have an impact not only on technical and environmental aspects, but also social, aesthetic, and psychological, the equation becomes much more complex and should also include qualitative aspects.
- Built-in knowledge: the knowledge that is necessary to assess the meaning of the percept sequence with respect to the performance criteria. Not all percepts that an agent receives are important to the current performance, and thus should not have an impact on the decision of what to do next. For this assessment, knowledge is necessary. The implication for interactive buildings, to a certain extent, is a model of self and should be maintained.

For the outset of this discussion, we pose that agent-based theory is well-suited to describe interactive buildings. Thus, the architectural discourse should include research and understanding of computer science, interaction design, and cognitive science. In this paper, we provide a concise critical reading of such sources. We claim that interaction narratives form an approach to unify such concepts in an architectural and productive way.

2. Interaction in Architecture until Contemporary Disruption

Since the late 1980s technologies have developed that make it possible to dynamically change building (components). Examples of such technologies are media facades, kinetic structures, ambient environments, smart homes, and so on. Early experimentation made use of these technologies in short loops, meaning a sensor set; a controller deciding on some action based on the sensor information and an actuator or device that displays or acts in some way. Media facades are the oldest examples of such systems (e.g., Zeitgallery by Christian Möller in Frankfurt, 1992 [6] (pp. 95). In this paper, there is not enough space to elaborate on the historical development of such systems. We can note however, that today the technological loops become more extended and complex with technologies such as Internet of Things, wearables, and cloud computing [7]. These technologies tend to converge, greatly expanding the potential of interactive architecture [8]. It means that it has become easier for dynamic building components to be aware of the environment, to be much more linked to extensive sensor networks and the Internet, and to better anticipate (multiple) user needs. The complexity of communication chains between many such components has led some researchers to call these systems ecosystems [9]. As the name ecosystem suggests, we are dealing with highly interconnected systems that need to be approached in a more organic way.

3. The Case for Interaction

Many interactive systems have been developed already, mostly as installations, art pieces, or additions to spaces. Much of this work is informed by technological opportunity and experimentation. Yet, a comprehensive approach to understand from an interaction perspective is largely missing. The most notable exceptions from the architectural point of view are Michael Fox [10], who has created several interactive buildings in the past two decades, and Robert Kronenburg [2], who investigates the fundamental aspect of change in architecture.

There is rising awareness from the field of Human-Computer Interaction (HCI) towards the field of architecture (most strongly defended by Malcolm McCullough's book Digital Ground [11]), but the two fields are still much apart [12–14]. Given the recent nature of the phenomenon of interactive architecture, there is no conclusive evidence yet that unambiguously states the (dis)advantages of interactive architecture. Some experimental work does provide us with clues about this, however. Schnädelbach and his team built a bio-feedback based interactive prototype, called ExoBuilding [15], and conducted several experiments while measuring physiological responses from users. They noted a positive effect on the users in physiological sense, while on the other hand most users found the explicit bio-feedback disturbing and unpleasant. Coyle et al. noted several systems for mental health interventions that have positive effects on people, for example online treatments, mobile support, therapeutic computer games, virtual and augmented reality exposure therapies, relational agents, and robotic companions [16]. Wouters et al. (2016) analyzed in more detail the spatial and social aspects dealing with successful engagement of people with an interactive installation [17]. They identified encounters, triggers, and activation loops as important mechanisms in establishing and sustaining interaction.

4. Limitation of Contemporary Interactive Systems

There are many terms used in the literature to describe various kinds of interactive systems for buildings. In fact, there is a confusing amount of terms, many of which are not clearly defined, overlapping, or similar in meaning. From passive buildings to interactive buildings we can see a gradient development of dynamic, responsive, and interactive buildings. From each step, dynamic to responsive to interactive, the building has an increasing amount of user involvement in the changes that occur in the building (Figure 1).

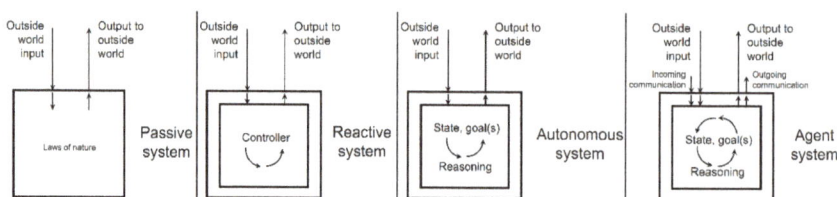

Figure 1. Increasing degrees of interaction in buildings.

In Figure 1 the box indicates the system; ingoing arrows are percepts and communications from the outside world; outgoing arrows are actions and communications to the outside world. Difference between systems can be described by their internal state and input, output types.

Passive systems arguably are not interactive at all; they are included to define the extreme end of the spectrum between non-interactive buildings and interactive buildings. Passive systems are mechanistically defined agents which respond to inputs from the outside world in predetermined ways. Reactive systems employ a simple controller system similar to "if X then Y" production systems. Autonomous systems record their own state and have a set of goals that they want to achieve. They use reasoning algorithms to select actions that minimize the distance between goal and current state. Agent systems can receive and send communications to other agents. Their reasoning algorithms are more complex and may also include utility-based prediction models that allow the agent to compare various future action sequences before actually deciding on the action it will perform in the outside world.

In research literature, we can find many different terms for responsive building, with each having a specific and different meanings. These terms are: Building Automation Systems, Smart Homes, sentient buildings, adaptive buildings, dynamic buildings, kinetic architecture, intelligent buildings, and portable buildings. In our research, we are particularly concerned with interactive buildings. The term interaction is often used to indicate anything that responds in some way to a kind of input. This very broad interpretation renders the term interaction quite meaningless and might as well indicate simple reactive systems such as automatic sliding doors. For this reason, we want to limit the use of the term interactive buildings only to such buildings that support a meaningful exchange of information between the user and the building, and the exchange influences changes in both the user and the building. Such levels of interaction today are not yet realized as buildings to the best of our knowledge, but mostly as installations or art pieces. Yet, we believe that the step towards such kinds of buildings is not far away in the future. Thus, we would do well to prepare a critical theoretical foundation for the design of such systems, which informs architects for the design of such buildings.

4.1. Social Aspects of Interactive Systems

A lot of research has been done on smart homes and intelligent environments with the major focus of this work on energetically optimal and sustainable well-performing buildings [18]. Surely sustainability is very important but there are strong clues that there are many more factors at play that would promote other behavior orientation than only energy performance. Walldén and Mäkinen [19] note that acceptance of smart environments depends not only on usefulness, but also on ease of use and trust, social influence, as well as on societal level cultural, economic, and legal factors. When we engage in an interactive exchange between user and system, the interaction takes on a deeper meaning than just pressing buttons. People will attribute personality traits to the system, so the system will be increasingly perceived as a social thing. In this sense, Tay et al. [20] observed that concerning social robot acceptance, attributed characters such as occupational role, gender, and personality of the robot play an important role. Partala and Saari [21] conclude that successful technology adoption depend on emotional design as much as functionality and usefulness. How this emotional design should be achieved remains an open question. From the work of the researchers above, we may infer that interactive systems need a variety of roles to best support the user.

Interaction has social implications as well. Mostly interactive installations are conceived between a system and one user. However, evidence from research shows that with the increasing number of people, the nature of the interaction changes as well. For example, the honeypot effect, [17] is the phenomenon where people who are engaged with a system stimulate by-passers to observe and ultimately engage with the system. Claes and Vande Moere [22] demonstrated identical displays compared in a public setting and isolated setting and how using a narrative and no narrative lead to a difference in comprehension and ease of use. Valkanova et al. demonstrated the impact of citizen-driven data visualization on perception, behavior change, social awareness, and public participation [23].

4.2. User Experience and Situatedness in Interactive Systems

In the field of interaction design, the concept of user experience is used to express a more user-grounded orientation of how systems and people may relate [24]. Although the concept of user experience is widely used in the field of Human-Computer Interaction, there is generally no accepted definition of user experience yet. There are marked differences in its conception and application based on geographical location and background [25]. What seems to play an important role in the success of an interactive system is the sense of locus of control [26]—meaning the degree to which an individual reflects about his/her capabilities to exert control in an environment. As a more concrete example, Meerbeek and his team [27] found many inadequacies in automated exterior blinds systems. They defined four different user profiles that perform better for the user (minimal, regular, active, and system control with manual override). The recognition of multiple user profiles is important, and leads to the (yet unanswered) question how to dynamically choose the proper user profile. Concerning the integration of interaction in the architectural design process, Houben et al. [28] claim that successful integration of interactive systems in architectural design projects can only be achieved when architects perceive said systems as a material that they can approach in much the same way they aim to express an architectural message.

Interactions in systems with multiple components quickly lead to complex relationships and decision situations that are difficult to predict. Beer demonstrated for a simple case (walking behavior of a legged agent under dynamic conditions) that the behavior of the agent is the result of the coupling of two dynamic systems—namely the agent and the environment—and cannot be assigned to just one of the systems [29]. This coupling is also known by the term situatedness, meaning that we cannot understand an agent's behavior unless we consider the way it is linked to its environment. Clancey additionally stated that much of what we do is a re-coordination of previous combinations of perceiving (or sensing), conceiving (or deciding), and moving (or acting) rather than the manipulation of an explicit knowledge model [30] (pp. 1–2).

For the design of interactive architecture this means that responsive buildings are by definition situated—which means that we cannot suffice with simple input–response schemes to guide the behavior of interactive buildings. With multiple components, also the concept of agency needs to be extended to the interaction of multiple agents together. Classical agent frameworks use utility as a measure to decide what to do. However, with multiple agents, and situatedness in the equation, utility values are not sufficient anymore. In multi-agent systems it is acknowledged that agents need to communicate and may not share the same utility- or agenda. Thus, agents need also to build assumptions about the other agents and assign levels of trust to agents in the system [31]. The most important implication, which cannot be stressed enough, is that we have to move beyond a utilitarian interpretation of performance maximization.

4.3. The Challenge for Architecture of Interactive Buildings

Based on the previous discussion we may conclude that there is no simple single solution that will work for complex systems like interactive buildings. Systems may have different goals: apart from performance, interactions may also be geared towards sustaining, servicing, symbolizing, and entertaining. Systems can engage in different styles with the user, such as in an instructive way, as a conversation, series of manipulations, or in an explorative way [32]. Mark Meagher stressed the "poetic potential" of responsive architecture, and notes that " . . . architects must develop a deep understanding of multiple types of change in buildings [33]." Cameline Bolbroe argued for a shift in attention away from the object to an "act of inhabitation," dealing with temporality, memory, learning, and emergence [34]. It seems evident that we need various interaction styles that an interactive building may adopt. This introduces the question how the building system would figure out which interaction style is the most appropriate, and how in an interactive manner the user different interaction styles may be adopted.

5. The Interaction Narrative

Settling on the proper interaction style with a user seems to be intimately bound with the user's experience, desires, and expectations. One approach that offers an integrated view of this is discussed by Maria Lehman [35]. Lehman's work is based in the domain of sensory design in healthcare environments. She notes that people's experiences are multi-sensory and that for a successful design it is necessary to connect well to the narrative of people in the building. In a healthcare situation, the narrative includes things as contemplation, visitors, sleep, recovery milestones, exercise, activities of daily living, medication, distraction, education, transition home, and pain management. A narrative, in other words, is a coherent story of the inhabitant, which needs to be supported by the activities or interactions of the building. In the more specific case of cancer treatment, Gillian Hayes and her team [36] have noted that " ... New technologies must accompany people on this journey while accommodating huge shifts in uses needs, motivations, energy levels and goals." We can generalize this to areas outside of healthcare. In 1999 Per Galle argued that a proper description of design should not be object-based but action-based—a notion which has strong links with the concept of narrative [37].

In our view the concept of narrative as described by Lehman is very relevant to interactive architecture. It has strong appeal because it enforces a consistent unfolding of events between the user and the building—thus it supports single interaction styles by making them clearer or readable for the user. Additionally, it enforces consistent reasons for role-switching between the user, the building, and the user-building relationship—thus it supports the decision process how to switch from one interaction style to another. The concept of narrative is very close to the notion of scenario, that is often used by architects to speculate about possible uses of their yet unrealized designs [38–41].

Narrative is usually associated with words and storytelling as can be readily seen in books and movies where the narrative is the prime structure. As such there is a very large amount of research on narrative in the written, spoken, and visual form. Our focus is on the role of narrative in technical systems, so we ignore narrative as a storytelling device by itself. Interactive narratives are stories, usually in computer games or installation art, where the user experiences a narrative through a storyline. Quite a lot of research and development has been invested in these kind of applications [42], although there is not much investigation into the user experience of such narratives [43]. Narratives have been advocated in computer system development as early as 1993 by Hasse Clausen [44]. Narratives are integrated today in the narrative of use cases [45], which has been the systematic approach to describe scenarios in software development since 1999 [46]. Moving closer to architectural design, Li-Shin Chang noted that a narrative does not need to be in the form of words but can contain objects as well, for example in landscape narratives [47]. Davidoff et al. observed in the context of control in smart homes for families, that just handing over control of the devices is not sufficient, but that the system should support families to control the things they value the most: "their time, their activities, and their relationship [48]."

Based on the discussion above, we propose that an interaction narrative is an organization of moments of interaction between the user and the system following a story that is consistent with an interaction style. Additionally, the interactive system has an interaction narrative for the way it switches between interaction styles—following a story that is consistent with user expectations.

With interactive architecture, we are fundamentally changing our understanding of buildings compared to almost all architectural thought of the past centuries, except for the earlier noted Cedric Price and Gordon Pask. Interaction narratives have the potential to unify technologies, aesthetical, and social aspects in a meaningful way. By respecting a narrative in the design process, it may be avoided that unbalanced attention goes to singular aspects of interaction, such as showcasing technology, or installations that do not deepen people's understanding of the built environment. It must be noted that the implications of this change are unclear. It will require an orchestrated effort from architects, researchers, legislators, clients, and people to advance our understanding.

Are interaction narratives the only solution for the development of responsive buildings? One is right to point out that narratives are possibly but one of the many options for such future development. Contemporary architectural theory, however, does not deal with the question, "how should my building behave?" Rather it focuses on performance connected to geometry and material, for example [49–51], or at best, agency as a technological and cultural phenomenon [52,53]. At the moment, narratives form the most developed theories that may account for a comprehensive way to design responsive architecture.

6. Conclusions

At the beginning of this article it was observed that buildings have a long time-span of use, and in this time-span must fulfill a dynamic range of functions. Passive technologies alone are capable of covering some of this range, but not everything. For reasons of functionality, comfort, safety, and pleasure, increasing amounts of flexible additions are made to the building outfit, i.e., dynamic shading systems, intelligent HVAC installations, automatic doors, and so on. Today, technological advances make it possible to link up building components to each other locally and globally, for example through IoT technology. This leads to a revised notion of buildings, in which they become increasingly actively engaged with the needs of the users. At some point such systems become so advanced that we start to call them in an uncritical manner responsive or interactive.

It is this uncritical manner of use of the term responsive or interactive that we investigated in this article. We hope to have demonstrated that the notion of responsive architecture is more complex than just endlessly adding sensors, actuators, and components. Adding such things is an act that deeply changes the building, not only at the technical level, but also at the cultural and social level. There are three main points derived from the work:

1. Agency: First of all, we must distance ourselves from the idea that a building is a passive thing to which we can add technology, and that by doing so, the building still is that unaffected passive thing. We must learn to see a building as an agent, one that interacts with other agents—be it other buildings, people, or software systems. Agent theory and multi-agent theory gives us the technical and formal tools by which we can create functional models of responsive buildings. Currently, agent theory is well-developed in the field of Artificial Intelligence for well over two decades, but the field of architecture has hardly touched upon this concept. This sets a theoretical agenda to incorporate the notion of agency in architecture.

2. Situatedness: Second, we must understand that with a richer arsenal of possible responses by the building, also the context and the history of responses start to play a decisive role. In other words, responsive systems are situated. As stated by John Gero, situatedness means that "where you are when you do what you do matters." [54] It is impossible, and even unnecessary, to try and capture all possible information flows for a responsive building. With the concept of situatedness, we can make an informed choice about what aspects of the context and history are taken into consideration by the responsive building. Making these choices explicit and integrating them into the reasoning process of the responsive building is an important step in the design of the building. This sets a theoretical agenda to formulate the proper selection of quantitative parameters from which a model of the context and history of the context can be built.

3. Narratives: Third, the situatedness thesis implies that the unfolding of a series of interaction events between agents is a meaningful act that develops in a logical way under a series of assumptions by the agents. Such a logical development of interaction events is a narrative, as recognized in the work by Maria Lehman. We pose that interaction narratives currently are the best viewpoint from which to develop responsive and interactive architecture. With the interaction narrative we inevitably move away from objective and quantifiable parameters that are part of agency and situatedness. From an engineering/technical point of view, this may raise the concern that narratives require too much subjective aspects to be actually useful in the design of responsive buildings. The concept of narrative, however, does not hide the subjective aspect

of design, but indeed makes it an explicit part of the design discourse and thus something that can be compared and judged, albeit in a subjective manner. This sets a theoretical agenda of how proper interaction narratives should be constructed, how they can be formulated using agency and situatedness, and how they can be evaluated.

To conclude, creating an interactive building should be more than the unrelated collection of many responsive components in a building. The notion of interaction narrative allows the design team of interactive systems to bring all possible moments of interaction into a coherent whole. Since a narrative contains a sequence of events, it also forces designers to consider user interactions as they should happen after each other, and how they could guide the user from event to event. Whereas the traditional way of architects to conceive the possible use of buildings is through a spatial sequence, responsive building design requires the integration of events and spaces into narratives for people. The people aspect has to be stressed because the design of narratives cannot be done in a meaningful manner without more explicit consideration of all the possible users in a building. The challenge therefore is to come up with a design framework for responsive buildings that is much richer and more complex than contemporary approaches. It is our opinion that much of the success of future responsive buildings will lie in an approach that combines events, spaces, and people.

As we are at the beginning of interactive buildings, a lot of work and experiments are still ahead of us. This position paper makes the case for interaction narratives as a promising future direction. Our claim here is theoretical, and that is obviously its main weakness. The real impact of interactive architecture cannot only be studied in a theoretical approach. Future work must deal with physical prototypes that confront reality and people, and which is assessed in the wild. Whether user narratives truly fulfill this potential, can only be found by prototyping, user testing, and implementing designs in real buildings.

Prototyping and building responsive buildings play a crucial role if we want to reach a more mature understanding of what responsive and interactive buildings may be. Theoretical observations alone are not enough as they are most likely to overlook aspects that could not be foreseen only theoretically. Additionally, since we are building types of systems that have not existed before, our understanding can only grow in a continuous cycle of reflection and action. However, without theoretical reflection, we risk that our explorations are technology-driven only, without asking ourselves what we really want. We hope that the current discussion in this position paper contributes to that question.

Funding: This research was partially funded by the grant CELSA, grant number CELSA/18/020.

Acknowledgments: Part of the research reported in this paper is supported by the grant CELSA/18/020, which is a collaboration between CTU Prague, Prague, Czech Republic and KU Leuven, Leuven, Belgium. The support of this grant is gratefully acknowledged.

Conflicts of Interest: The authors declare no conflict of interest.

References

1. Leupen, B.; Heijne, R.; van Zwol, J. *Time-Based Architecture*; 010 Publishers: Rotterdam, The Netherlands, 2005.
2. Kronenburg, R. *Flexible: Architecture that Responds to Change*; Laurence King Publishing: London, UK, 2007.
3. Pask, G. Architectural relevance of cybernetics. *Archit. Des.* **1969**, *39*, 494–496.
4. Achten, H. One and Many: An Agent Perspective on Interactive Architecture. In *ACADIA 2014 Design Agency, Proceedings Proceedings of the 34th Annual Conference of the Association for Computer Aided Design in Architecture, Los Angeles, CA, USA, 23–25 October 2014*; Gerber, D., Huang, A., Sanchez, J., Eds.; Riverside Architectural Press: Toronto, Canada, 2014; pp. 479–486.
5. Russell, S.; Norvig, P. *Artificial Intelligence: A Modern Approach*, 3rd ed.; Prentice Hall: Upper Saddle River, NJ, USA, 2010.
6. Mignonneau, L.; Sommerer, C. Media Facades as Architectural Interfaces. In *The Art and Science of Interface and Interaction Design*; Sommerer, C., Jain, L.C., Mignonneau, L., Eds.; Springer: Berlin/Heidelberg, Germany, 2008; Volume 1, pp. 93–104.

7. Achten, H. Closing the Loop for Interactive Architecture-Internet of Things, Cloud Computing, and Wearables. In *Real Time, Proceedings of the 33rd eCAADe Conference, Vienna, Austria, 16–18 September 2015*; Martens, B., Wurzer, G., Grasl, T., Lorenz, W.E., Schaffranek, R., Eds.; Vienna University of Technology: Vienna, Austria, 2015; Volume 2, pp. 623–632.

8. Hansen, K.; McLeish, T. Designing Real Time Sense and Response Environments through UX Research. In *Real Time, Proceedings of the 33rd eCAADe Conference Vienna, Austria, 16–18 September 2015*; Martens, B., Wurzer, G., Grasl, T., Lorenz, W.E., Schaffranek, R., Eds.; Vienna University of Technology: Vienna, Austria, 2015; Volume 2, pp. 651–658.

9. Crowley, J.L.; Coutaz, J. An ecological view of Smart Home technologies. In *Ambient Intelligence. Lecture Notes in Computer Science 9425*; De Ruyter, B., Kameas, A., Chatzimisios, P., Mavrommati, I., Eds.; Springer: Berlin/Heidelberg, Germany, 2015; pp. 1–16.

10. Fox, M.; Kemp, M. *Interactive Architecture*; Princeton Architectural Press: New York, NY, USA, 2009.

11. McCullough, M. *Digital Ground: Architecture, Pervasive Computing, and Environmental Knowledge*; The MIT Press: Cambridge, MA, USA, 2004.

12. Dalton, N.; Green, K.; Marshall, P.; Dalton, R.; Hoelscher, C.; Mathew, A.; Kortuem, G.; Varoudis, T. Ar-CHI-Tecture: Architecture and Interaction. In *CHI EA '12 CHI '12 Extended Abstracts on Human Factors in Computing Systems*; ACM: New York, NY, USA, 2012; pp. 2743–2746.

13. Dalton, N.; Green, K.; Dalton, R.; Wiberg, M.; Hoelscher, C.; Mathew, A.; Schnädelbach, H.; Varoudis, T. Interaction and Architectural Space. In *CHI EA '14 CHI '14 Extended Abstracts on Human Factors in Computing Systems*; ACM: New York, NY, USA, 2014; pp. 29–32.

14. Salem, F.; Haque, U. Urban Computing in the Wild: A survey on large scale participation and citizen engagement with ubiquitous computing, cyber physical systems, and Internet of Things. *Int. J. Hum. Comput. Stud.* **2015**, *81*, 31–48. [CrossRef]

15. Schnädelbach, H.; Irune, A.; Kirk, D.; Glover, K.; Brundell, P. ExoBuilding: Physiologically driven adaptive architecture. *ACM Trans. Comput. Hum. Interact.* **2012**, *19*, 25. [CrossRef]

16. Coyle, D.; Thieme, A.; Linehan, C.; Balaam, M.; Wallace, J.; Lindley, S. Preface Emotional Wellbeing. *Int. J. Hum. Comput. Stud.* **2014**, *72*, 627–628. [CrossRef]

17. Wouters, N.; Downs, J.; Harrop, M.; Cox, T.; Oliveira, E.; Webber, S.; Vetere, F.; Vande Moere, A. Uncovering the Honeypot Effect: How Audiences Engage with Public Interactive Systems. In *Proceedings of the 2016 ACM Conference on Designing Interactive Systems (DIS '16)*; ACM: New York, NY, USA, 2016; pp. 5–16. [CrossRef]

18. Augusto, J.C. Past, Present and Future of Ambient Intelligence and Smart Environments. In *Agents and Artificial Intelligence. Communications in Computer and Information Science 67*; Filipe, J., Fred, A., Sharp, B., Eds.; Springer: Berlin/Heidelberg, Germany, 2010; pp. 3–15.

19. Walldén, S.; Mäkinen, E. On accepting smart environments at user and societal levels. *Univers. Inf. Soc.* **2014**, *13*, 449–469. [CrossRef]

20. Tay, B.; Jung, Y.; Park, T. When stereotypes meet robots: The double-edge sword of robot gender and personality in human-robot interaction. *Comput. Hum. Behav.* **2014**, *38*, 75–84. [CrossRef]

21. Partala, T.; Saari, T. Understanding the most influential user experiences in successful and unsuccessful technology adoptions. *Comput. Hum. Behav.* **2015**, *53*, 381–395. [CrossRef]

22. Claes, S.; Vande Moere, A. The impact of a narrative design strategy for information visualization on a public display. In *Proceedings of the 2017 Conference on Designing Interactive Systems (DIS 2017), Edinburgh, UK, 10–14 June 2017*; ACM: New York, NY, USA, 2017; pp. 833–838.

23. Valkanova, N.; Jorda, S.; Moere, A.V. Public visualization displays of citizen data: Design, impact and implications. *Int. J. Hum. Comput. Stud.* **2015**, *81*, 4–16. [CrossRef]

24. Sharp, H.; Rogers, Y.; Preece, J. *Interaction Design: Beyond Human-Computer Interaction*, 2nd ed.; John Wiley & Sons: Chichester, NH, USA, 2007.

25. Lallemand, C.; Gronier, G.; Koenig, V. User experience: A concept without consensus? Exploring practitioners' perspectives through an international survey. *Comput. Hum. Behav.* **2015**, *43*, 35–48. [CrossRef]

26. Jang, J.; Shin, H.; Aum, H.; Kim, M.; Kim, J. Application of experiential locus of control to understand users' judgments towards useful experience. *Comput. Hum. Behav.* **2016**, *54*, 326–340. [CrossRef]

27. Meerbeek, B.; Kulve, M.T.; Gritti, T.; Aerts, M.; Van Loenen, E.; Aarts, E. Building automation and perceived control: A field study on motorized exterior blinds in Dutch offices. *Build. Environ.* **2014**, *79*, 66–77. [CrossRef]

28. Houben, M.; Denef, B.; Mattelaer, M.; Claes, S.; Vande Moere, A. The Meaningful Integration of Interactive Media in Architecture. In *Proceedings of the 2017 ACM Conference Companion Publication Conference on Designing Interactive Systems (DIS '17 Companion Volume), Edinburgh, UK, 10–14 June 2017*; ACM: New York, USA, 2017; pp. 187–191.

29. Beer, R.D. A Dynamical Systems Perspective on Agent-Environment Interaction. In *Artificial Intelligence: Critical Concepts Volume III*; Chrisley, R., Ed.; Routledge: London, UK, 2000; pp. 210–255.

30. Clancey, W.J. *Situated Cognition: On Human Knowledge and Computer Representations*; Cambridge University Press: Cambridge, UK, 1997.

31. Weiss, G. (Ed.) *Multiagent Systems*; The MIT Press: Cambridge, MA, USA, 2001.

32. Achten, H. Buildings with an Attitude: Personality Traits for the Design of Interactive Architecture. In *Faculty of Architecture, Computation and Performance, Proceedings of the 31st eCAADe Conference (Volume 1), Delft, The Netherlands, 18–20 September 2013*; Stouffs, R., Sariyildiz, S., Eds.; Delft University of Technology: Delft, The Netherlands, 2013; pp. 477–485.

33. Meagher, M. Designing for change: The poetic potential of responsive architecture. *Front. Archit. Res.* **2015**, *4*, 159–165. [CrossRef]

34. Bolbroe, C. Mapping the intangible: On adaptivity and relational prototyping in architectural design. In *Architecture and Interaction. Human Computer Interaction in Space and Place*; Dalton, N.S., Schnädelbach, H., Wiberg, M., Varoudis, T., Eds.; Springer International Publishing: Basel, Switzerland, 2016; pp. 205–229.

35. Lehman, M.L. Environmental Sensory Design. In *Intelligent Buildings: Design, Management, and Operation*, 2nd ed.; Clements-Croome, D., Ed.; ICE Publishing: London, UK, 2013; pp. 61–70.

36. Hayes, G.R.; Abowd, G.D.; Davis, J.S.; Blount, M.L.; Ebling, M.; Mynatt, E.D. Opportunities for Pervasive Computing in Chronic Cancer Care. In *Pervasive Computing. Pervasive 2008. Lecture Notes in Computer Science*; Indulska, J., Patterson, D.J., Rodden, T., Ott, M., Eds.; Springer: Berlin/Heidelberg, Germany, 2008; Volume 5013, pp. 262–279.

37. Galle, P. Design as intentional action: A conceptual analysis. *Des. Stud.* **1999**, *20*, 57–81. [CrossRef]

38. Carp, J. Design participation: New roles, new tools. *Des. Stud.* **1986**, *7*, 125–132. [CrossRef]

39. Hasdogan, G. The role of user models in product design for assessment of user needs. *Des. Stud.* **1996**, *17*, 19–33. [CrossRef]

40. Uluoglu, B. Design knowledge communicated in studio critiques. *Des. Stud.* **2000**, *21*, 33–58. [CrossRef]

41. Crilly, N.; Cardoso, C. Where next for research on fixation, inspiration and creativity in design? *Des. Stud.* **2017**, *50*, 1–38. [CrossRef]

42. Ben-Arie, U. The Narrative-Communication Structure in Interactive Narrative Works. In *Interactive Storytelling. Lecture Notes in Computer Science 5915*; Iurgel, I.A., Zagalo, N., Petta, P., Eds.; Springer: Berlin/Heidelberg, Germany, 2008; pp. 152–162.

43. Milam, D.; El-Nasr, M.S.; Wakkary, R. Looking at the interactive narrative experience through the eyes of the participants. In *Interactive Storytelling, First Joint International Conference on Interactive Digital Storytelling*; Springer: Berlin, Germany, 2008; pp. 96–107.

44. Clausen, H. Narratives as tools for the system designer. *Des. Stud.* **1993**, *14*, 283–298. [CrossRef]

45. Tena, S.; Díez, D.; Díaz, P.; Aedo, I. Standardizing the narrative of use cases: A controlled vocabulary of web user tasks. *Inf. Softw. Technol.* **2013**, *55*, 1580–1589. [CrossRef]

46. Booch, G.; Rumbaugh, J.; Jacobson, I. *The Unified Modeling Language User Guide*; Addison-Wesley: Reading, MA, USA, 1999.

47. Chang, L.-S.; Bisgrove, R.J.; Liao, M.-Y. Improving educational functions in botanic gardens by employing landscape narratives. *Landsc. Urban Plan.* **2008**, *86*, 233–247. [CrossRef]

48. Davidoff, S.; Lee, M.K.; Yiu, C.; Zimmerman, J.; Dey, A. *UbiComp 2006: Ubiquitous Computing, Lecture Notes in Computer Science 4206*; Dourish, P., Friday, A., Eds.; Springer: Berlin/Heidelberg, Germany, 2006; pp. 19–34.

49. Kolarevic, B.; Malkawi, A.M. *Performative Architecture: Beyond Instrumentality*; Spon Press: New York, NY, USA, 2005.

50. Lynn, G. *Composites, Surfaces, and Software: High Performance Architecture*; Yale School of Architecture: New Haven, CT, USA, 2010.

51. Berkel, B.; van Bos, C. *Knowledge Matters*; Frame Publishers: Amsterdam, The Netherlands, 2016.
52. Gerber, D.J.; Ibañez, M. (Eds.) *Paradigms in Computing: Making, Machines, and Models for Design Agency in Architecture*; eVolo Press: Los Angeles, CA, USA, 2014.
53. Velikov, K.; Thün, G. Towards an Architecture of Cyber-Physical Systems. In *Paradigms in Computing: Making, Machines, and Models for Design Agency in Architecture*; Gerber, D.J., Ibañez, M., Eds.; eVolo Press: Los Angeles, CA, USA, 2014; pp. 331–347.
54. Gero, J.S.; Kannengiesser, U. The Situated Function-Behaviour-Structure Framework. In *Artificial Intelligence in Design'02*; Gero, J.S., Ed.; Kluwer Academic Publishers: Dordrecht, The Netherlands, 2002.

buildings

MDPI

Article

Rationalized Algorithmic-Aided Shaping a Responsive Curvilinear Steel Bar Structure

Jolanta Dzwierzynska

Department of Architectural Design and Engineering Graphics, Faculty of Civil and Environmental Engineering and Architecture, Rzeszow University of Technology, al. Powstancow Warszawy 12, 35-959 Rzeszow, Poland; joladz@prz.edu.pl; Tel.: +48-17-865-1507

Received: 22 February 2019; Accepted: 7 March 2019; Published: 11 March 2019

Abstract: The correlation of the architectural form and the structural system should be the basis for rational shaping. This paper presents algorithmic-aided shaping curvilinear steel bar structures for roofs, using modern digital tools, working in the environment of Rhinoceros 3D. The proposed method consists of placing the structural nodes of the shaped bar structure on the so-called base surface. As the base surface, the minimal surfaces with favorable mechanical properties were used. These surfaces were obtained in two optimization methods, due to both the structural and functional requirements. One of the methods used was the so-called form-finding method. It wasalso analyzed the amount of shadow produced by the roof and the adjacent building complex, during a certain research period, to find the roof's optimal shape. The structure of the optimal shape was then subjected to structural analysis and its members were dimensioned. The dimensioning was carried out for two bar cross-sections, and as the optimization criterion, the smallest structure's mass was used. The presented research aims to show how it is possible to use generative shaping tools, so as not to block the creative process, to obtain effective, responsive structural forms, that meet both architectural and structural requirements.

Keywords: steel bar structures; structural analysis; parametric design; algorithmic-aided shaping; responsive architecture; shadows; Grasshopper; Karamba 3D

1. Introduction

Over the years, various methodologies have been developed for shaping building objects in accordance with the current development of design ideas, building materials, as well as technologies. The basic principles of composition in architecture and construction were set by Vitruvius in The Ten Books of Architecture, in the early centuries [1]. According to them, architecture should be based on a combination of three elements: stability, utility and beauty. Classical architecture was mainly based on the orderly composition of shapes formed mainly on the basis of Platonic solids [2]. In turn, nonlinear and organic architecture was the result of studying biological organisms, the structure of matter, and the application of this knowledge to design and construction. On the other hand, according to the dominant concepts of modernism, the shaping of structures was based on industrial technologies, functionalism and universal models, as well as modern ways of construction and the use of new materials [2]. That is why, over the years, the process of shaping building objects has been changing depending on the contemporary canons of beauty, materials, as well as the technological conditions. However, irrespective of these conditions, the process of shaping any building object always has to be adapted to both architectural and structural requirements, which are interrelated. Generally, the shaping process can be defined as the optimization of the building's shape and form, so that it meets the assumed initial criteria to the greatest extent. This means looking for such shapes and dimensions of the shaped building object, that would allow it to meet the requirements resulting from its purpose, as well as its future use. Requirements of the contemporary design standards for structures, as well as requirements of building law, are comprehensive. They are the set of related conditions regarding reliability, the load-bearing capacity, serviceability, durability, the resistance to exceptional impacts, and harmful impacts on the natural environment and society, etc. [3–8].

Due to this fact, the shaping of any architectural/structural spatial system can be considered in various aspects, depending on the designing phase, as well as the nature and complexity of the design task. However, the shaping criteria always result directly from the starting boundary conditions for shaping: The structure's function, safety requirements, as well as the material and technological solutions.

The shaping phase is the phase preceding all subsequent stages of the design process, which is why it is the most creative phase on the one hand, and on the other hand, has a significant impact on the final form of the object [9,10]. Therefore, it very important that as many aspects as possible concerning the future project be considered in the initial phase. It can guarantee the creation of a sensitive, adaptable, and sustainable form, as well as a reliable structure. In addition, the conceptual phase is related to the multi-variant analysis of initial solutions, most often undertaken within a limited time. These conditions require the use of comprehensive tools that define the structural model, as well as accounting for the team and interdisciplinary nature of the designing [11–13].

Due to this fact, another factor influencing the way of shaping is the type and availability of design tools, which change with the development of technology [13–15]. Especially, during the last twenty years, the advancement of digital technologies has influenced the whole field of engineering design, and digital media have become a convenient means for shaping structures.

Various Computer Aided Design CAD tools enabled not only the creation of two-dimensional documentation [16], but also the creation of three-dimensional models based on two-dimensional drawings [17–19]. Moreover, thanks to the widespread use of digital tools in designing, the boundary between physical and virtual models have blurred. This was mostly caused by the hybridization of several methods and techniques for acquiring the model's geometry, as well as the development of reverse engineering [20]. Further, building information modeling (BIM), as a 3D model-based approach, gave architecture and civil engineering possibilities to streamline the design process by improving communication between the participants of the design process.

Especially, the development of the concepts and software enabling smooth modeling, which is digital modeling based on the non-uniform rational B-spline (NURBS) and its application in architecture, resulted in a significant change of the shaping process [21]. This was caused by the ability to control the geometry of shaped forms on an ongoing basis, enabling the creation of dynamic, parametric and non-linear forms. In turn, flexible geometry initiated the development of parametric and respond forms, which can evolve during shaping. Such possibilities created modern software tools like Computer Aided Three-dimensional Interactive Application CATIA, as well as Rhinoceros 3D and Dynamo, etc. equipped with visual scripting languages, which have revolutionized the shaping process. Today, parametric shaping of building structures allows one to create complex shapes, as well as their fabrication. Virtual spatial models of compound forms are created by means of advanced algorithms, based on parametric equations that designers can adjust for particular circumstances. Specialized computer software can generate original and atypical spatial structures, with mathematical precision and optimization [22–25]. Along with this line of thought, the paper discusses a novel algorithmic-aided approach to shaping curvilinear steel bar roof structures with the application of design tools, working in the environment of Rhinoceros 3D.

Curvilinear steel bar structures are defined as spatial structures made of slender members, directly connected bars, which carry loads. Historically, curvilinear steel bar structures, mostly in the form of lattice cylindrical structures, began to be applied in the mid-nineteenth century. However, due to serious difficulties in their calculation and construction because of the repeatable elements, they began to be used on a larger scale only in the 1960s.

However, the inherent properties of steel, such as its high strength-to-weight ratio, make the range of the use of steel bar structures very wide. They can be used as load-bearing structures of various roofs, as well as vertical and horizontal partitions of different topologies, in order to separate the space of various buildings, which also minimizes the visual mass of each structure.

Nowadays, their use is increasing, thanks to advanced technologies. Therefore, more and more curvilinear steel bar structures are created in a great variety of geometric forms and technical solutions, that constitute original and sometimes amazing examples of engineering inventions.

The research presented in this manuscript is focused on the rational parametric shaping of responsive curvilinear steel bar structures of a building, covering over a marketplace. This process is recognized as a process, where the aim is to shape a structure that satisfies a predetermined geometry and functionality, with respect to structural demands, as well as environmental conditions. In this research algorithmicaided shaping curvilinear steel bar structures is done by application of structural model making algorithms.

The algorithmic-aided structural shaping presented in the paper, where both the geometric model and structural analysis are realized by means of algorithms, is a rather new field of research. However, some researchers explore the concept of algorithmic-aided shaping in architecture, engineering and construction [26–29].

2. Materials and Methods

The task was to shape a covering of a marketplace, between two buildings, the orthogonal projection of which, is a rectangular trapeze (Figure 1). The dimensions of the trapeze were as follows, the lengths of the trapezoid bases: 14 m, 20 m, the length of the trapezoid height: 14 m.

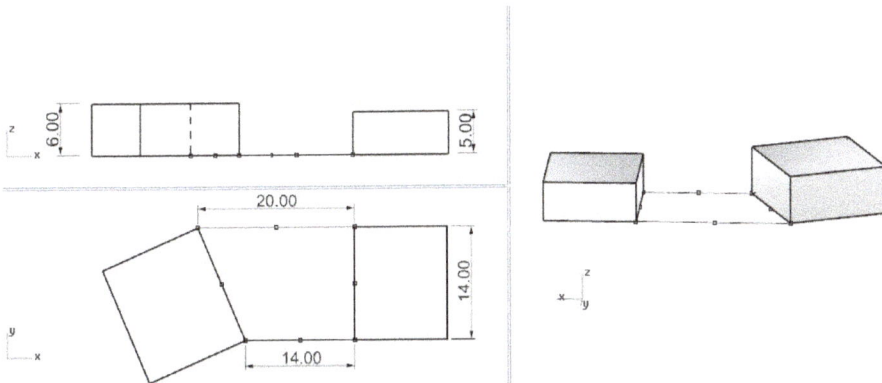

Figure 1. Horizontal, vertical and axonometric view of the considered square with the adjacent buildings.

The roof structure was shaped as a curvilinear steel two-layered bar structure. As a roofing material, polycarbonate plastic sheets were used, as well as tubes for structural elements.

In order to obtain the optimal roof structure, the roof shaping was carried out using modern tools for algorithmic-aided design, working in the environment of Rhinoceros 3D (Robert McNeel & Associates). The reason for which the above-mentioned tools were selected, is that they gave the possibility to optimize the model with the assumptions of the various initial criteria.

The approach to algorithmic-aided shaping of curvilinear steel bar structures of the roof proposed in this research, was realized by application of versatile tools: Grasshopper, Karamba 3D, and Ladybug, working in the mentioned modeling software, Rhinoceros 3D. We also used advanced software for structural analysis in order to optimize the bars' cross sections: Autodesk Robot Structural Analysis Professional 2019 [30].

In order to generate a parametric model, the educational version of Rhinoceros 5.0 in combination with the parametric tool Grasshopper, was used. Rhinoceros 3D is three-dimensional modeling software, which enables creation of various free form shapes, based on NURBS surfaces. Whereas, its plug-in, Grasshopper, enables the creation of scripts to describe parametric models, which can be modified and visualized in Rhino's viewport and carefully analyzed. However, Grasshopper's plug-in, Ladybug, was used for the shadow range analysis.

Integration of geometric shaping and structural analysis was carried out using Karamba 3D. This plugin made it possible to combine parameterized complex geometric forms of various topologies, load calculations and perform finite element analysis, according to Eurocode 3. Karamba 3D is intended to be a fast tool for structural optimization [31]. It has been used mostly in the early stage of design, although it is not limited to it. However, in the presented research, in order to perform structural analysis of complex curvilinear steel bar structures, the advanced professional software Robot Structural Analysis Professional, was used.

The validation of the functionality of the shaping tools mentioned above, as well as the benefits of their application, were accomplished by implementing them for the shaping and analysis of the responsive curvilinear steel bar structure.

The shaping strategy applied in the research consisted of the shaping of curvilinear steel bar structures by placing the structural nodes on the so-called base surface [29]. The base surface was a minimal surface obtained as a result of the optimization, or a surface received by the method of form-finding.

3. Results of Shaping a Roof Based on a Minimal Surface

According to mathematical definition, a minimal surface is a surface that locally minimizes its area, which is equivalent to having zero mean curvature. It is a surface with the smallest possible area among all surfaces stretched on given lines.

The application of minimal surfaces in architecture is not new. They are used mostly in the construction of light roof structures. Moreover, due to their characteristics, there is a greater and greater interest in the study of minimal surfaces and their wider use to create innovative architectural objects. Moreover, the application of minimal surfaces brings measurable benefits in structural design since the minimal surfaces exhibit an optimal system of forces and stresses. It is also an important task from an architectural and economic point of view, as the application of minimal surfaces means the use of a minimal area of claddings and the minimization of costs. Due to this fact, the parametric study and optimization were conducted on the examples of spatial lattices, created based on minimal surfaces.

3.1. Shaping Free-Form Roof Due to Optimization of the Minimal Surface

The first step in shaping the roof was formation of a minimal base surface. The minimal surface was determined as a surface stretched on four curved lines lying in vertical planes passing through the sides of the trapeze. Each line was established parametrically by three characteristic points, which could change their positions. Two of the points were the arches' ends and the third point was an inner point of the arch. The rectangular projections of these points on the horizontal plane were, respectively, the square's vertices and the centers of the square's sides (Figure 2).

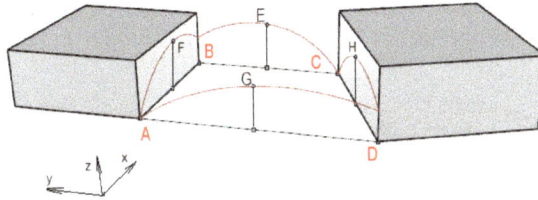

Figure 2. Establishing assumed arches for creation of a roof's base surface.

The position of characteristic points: A, B, C, D, E, F, G, H, were variables that could take different values, and were defined as local increments of the co-ordinates x, y, z. The intervals of these variables were defined for individual points as follows:

- Point A, B, C, D – z: 0.0 m–2.5 m.
- Point F, E, G – z: 4.0 m–5.0 m.
- Point E: x: 1.0 m–2.0 m (according to direction of axis x).
- Point G: x: 1.0 m–2.0 m (in the opposite direction than the axis x), Figure 2.

Depending on the adopted variables, we could get different shapes of arcs, and therefore, different shapes of the minimal surfaces defined by them. Consequently, this led to obtaining many alternative base surfaces. Some of them are presented in Figure 3.

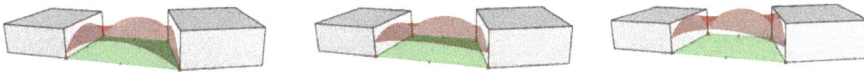

Figure 3. Generations of possible solutions of covering surfaces.

The optimal surface selected from the possible alternatives, was the surface with the smallest area, as it could guarantee the minimum amount of materials used. In order to find such a surface, the optimization process was performed in Grasshopper environment. As the optimization criterion, the minimum surface area was assumed, while the variable parameters were the individual positions of the points. Due to the optimization carried out, a minimal surface area equal to 263.042 m^2 was obtained for the surface, as well as characteristic heights of the points (coordinate z), presented in Figure 4. Additionally, both points, E and G, were moved horizontally one meter outside the trapeze bases, according to axis x.

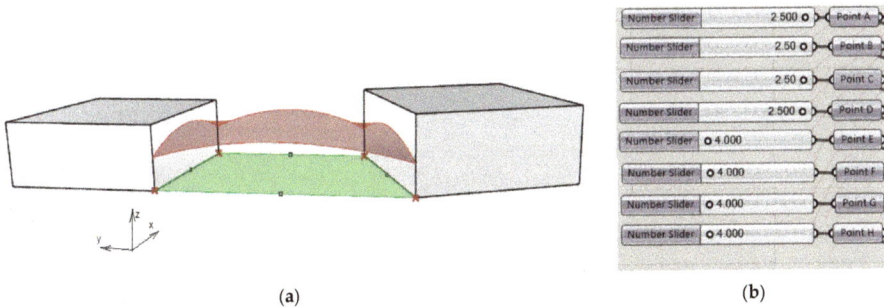

(a)   (b)

Figure 4. The best result of the roof surface optimization: (**a**) Perspective view of the roof's base surface; (**b**) The heights in meters of the characteristic roof's points.

This optimal surface constituted a base surface in order to create the roof's steel bar structure. The multi-layered grid on this surface was applied in such a way that the structural nodes of the top layer were included in the surface. In order to distribute the nodes, the surface was divided into the same number of segments, equal to ten in the x and y directions (Figures 4 and 5).

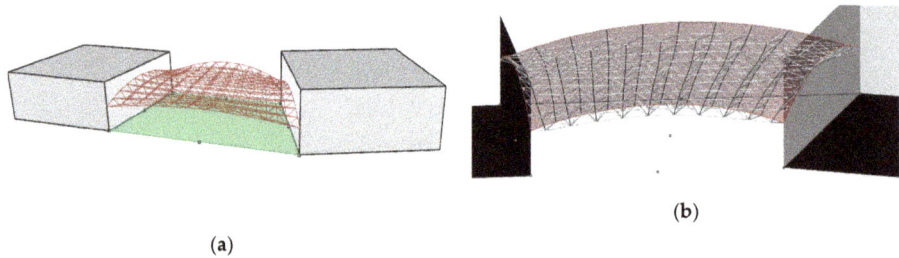

(b)

(a)

Figure 5. The roof structure received due to optimization: (**a**) View from above; (**b**) view from below.

The orthogonal projections with dimensions of the considered structure are shown in Figure 6.

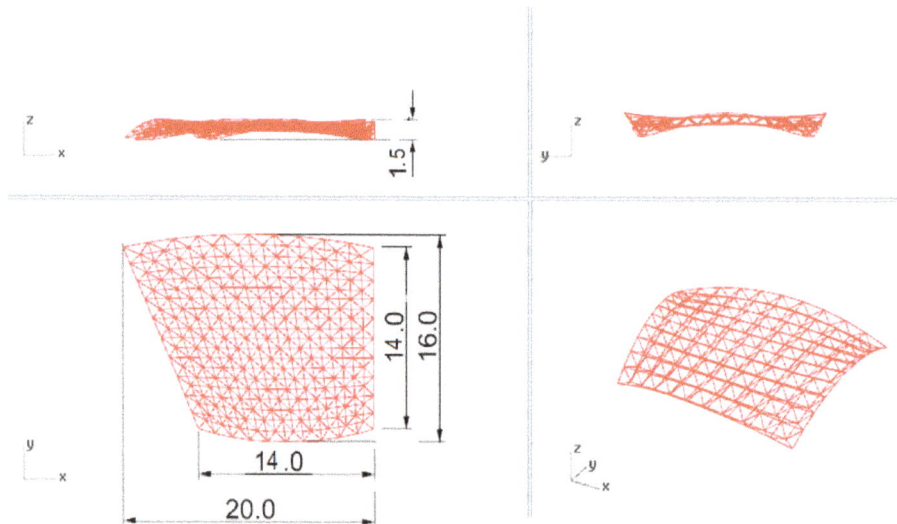

Figure 6. The views of the considered curvilinear steel bar structure.

3.2. Finding the Roof's Shape by the Form-Finding Method

Another way to shape the minimal base surface and the roof structure is using the so-called form-finding method. Historically, it is an quite an old process, although the tools for the realization of it have been developing over the years. There are two form-finding techniques: the first one it is a method, which uses hanging models to simulate compressive forces; and another one is the method of the minimal energy shapes of soap films. The hanging model method can be applied to simulate the behavior of a family of structures known as funicular structures. Thus, finding the shape of the structure based on the hanging model method means that the form found is adjusted to magnitudes and positions of the forces acting on it. This method is mostly used in order to shape structures, which work mainly in tension and compression.

The form-finding method has become very popular in recent years, mainly as a method for the shaping of free-form structures. Nowadays, due to the development of modern computerized modeling and calculating tools, which give the freedom to explore the design space, it can be realized automatically in various ways [32]. In the case of parametric shaping structures, form-finding is determined as the process in which parameters are directly controlled in order to find an optimal geometry of a structure, which is to be in static equilibrium with assumed loads. The load used for form-finding is usually the structure's self-weight, however, they can also be other external loads.

In the considered case, the roof's base surface was generated as the form found over the trapeze-shaped place, by means of plug-ins working in environment of Rhinoceros 3D. During such a form-finding process, the structural load direction had been inverted in order to achieve a proper arrangement of the shaped structure. The height of the roof during simulation was assumed to be the same as in the previous case, which was 1.5 m. On the generated surface, a two-layered grid structure with the same division along axes x and y, like in the previous case, was applied (Figure 7). The structure was placed at the same height as the structure presented in Section 3.1.

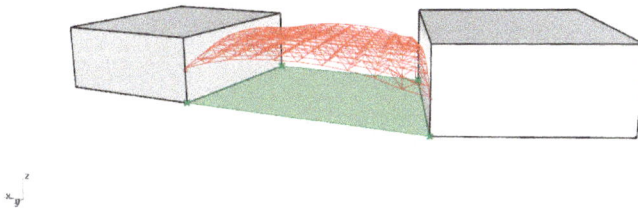

Figure 7. The roof structure generated due to form-finding process.

The orthogonal projections with dimensions of the obtained structure are shown in Figure 8.

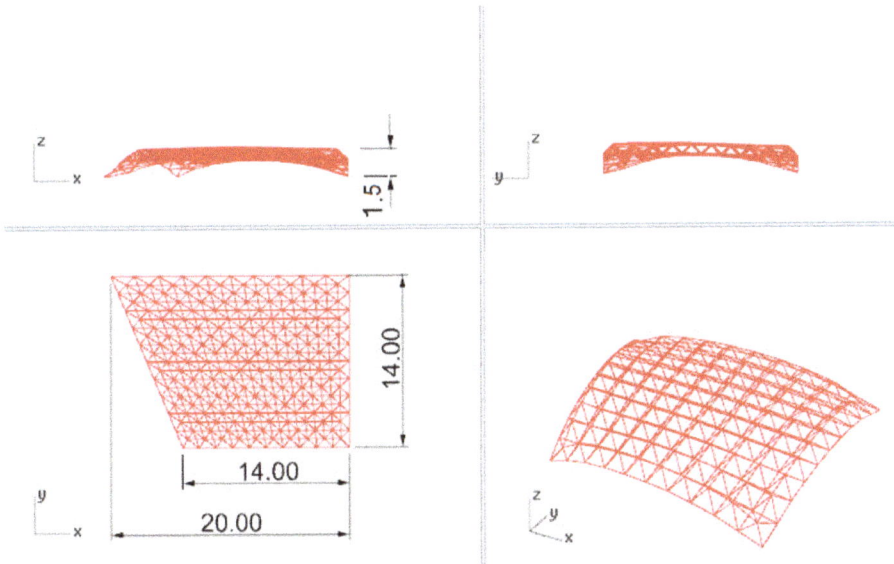

Figure 8. The views of the considered curvilinear steel bar structure.

3.3. Analysis of the Shadow Created by a Set of Connected Buildings

Due to the fact that the function of the shaped roof was not only to provide protection against atmospheric precipitation but also against the sun in the summer period, the shadow produced by the roof, together with the adjacent complex of buildings, was analyzed. The aspect of the construction of the shadow cast by the building complex is presented in [33]. In our research, the analysis was carried out for the period from March to September, during the hours 10 am to 5 pm by Grasshopper's plug-in, Ladybug. For the analysis, the city of Warsaw was assumed as the location of the building complex and the complex's position was with respect to the north direction, as shown in Figure 9. The analysis showed that the complex of buildings with the shaped roof, presented in Figure 5, produced much more shadow during the considered period, than the complex presented in Figure 7. The amount of shadow generated in the analyzed period is shown graphically in Figure 9.

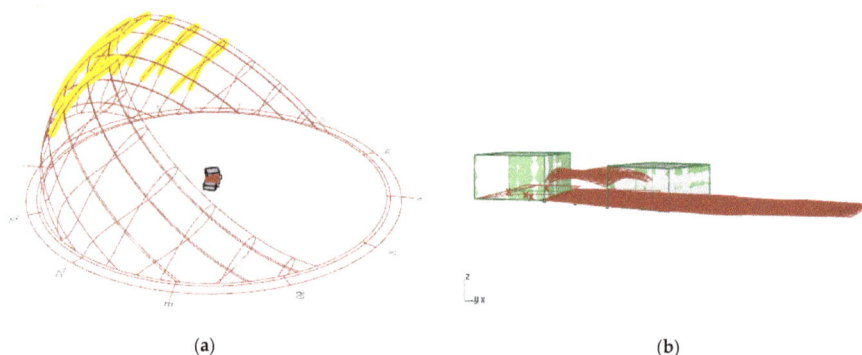

(a)   (b)

Figure 9. The analysis of shadows: (**a**) Location of the buildings with respect to the north direction; (**b**) the amount of shadow generated in the analyzed period.

3.4. Optimization of the Structural Members' Cross Sections

We carried out structural optimization using the Autodesk Robot Structural Analysis Professional 2019 software. The border conditions regarding the wind and snow zones were assumed according to the location of the building objects. Due to the shape of the roof, that is, its mostly flat nature, the snow load was assumed to be the same as for the shed roof. However, due to the roof's installation, that is, its adherence to adjacent buildings, the possibility of a snow drift has been included in the calculations [34]. The maps on bars showing the distribution of the axial force Fx are presented in Figure 10.

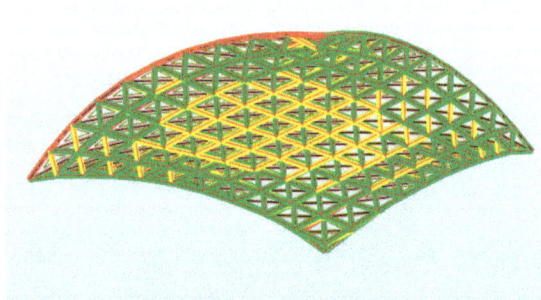

Figure 10. Maps on bars showing the distribution of axial force Fx.

The bars of the structure were divided into three groups for the dimensioning: The top truss bars, the bottom truss bars, and the truss diagonal bars. The structure was optimized assuming as the optimization criterion the mass of the structure. Moreover, the structure was optimized for two cross-sections: circular and square. The results of the optimization are presented in Table 1.

Table 1. The results of the structural optimization of the considered structure.

Kind of Member	Cross Section Box Hollow	Cross Section Circular Hollow
Top lattice's bars	$40 \times 40 \times 4$	38×3.6
Bottom lattice's bars	$40 \times 40 \times 4$	38×3.6
Diagonal bars	$40 \times 40 \times 4$	31.8×45

The considered structure was composed of 221 nodes and 800 members. However, the total mass of the shaped structure was equal to 186,525 kg in the case of members with box hollow cross-sections and 185,621.286, in the case of round tube members. Due to this fact as the most efficient structure, it has been chosen the structure with circular hollow cross-sections as the lighter structure than the structuer with box hollow cross sections.

4. Discussion

The use of Rhino and Grasshopper gave much creative freedom and flexibility in shaping. Due to this fact, it was very convenient to use during the initial design process. In addition, to the advantages of the possibility of the creation of many alternatives of geometric forms by Grasshopper, there was a huge potential for combining parametric shaping with interactive evolutionary optimization.

The conducted research on shaping the roof over the market square, using algorithmic-aided shaping, has shown that a comprehensiveapproach to shaping is possible. Therefore, as early as possible, we could take into account many aspects that affect the future form and work of the structure. These aspects can be conditions regarding the planned function, reliability, load-bearing capacity, resistance to exceptional impacts and natural environment.

Moreover, defining the geometry of the shaped structure by means of the algorithms, gave the possibility of various modifications and obtaining a virtually unlimited number of geometric forms. Especially, in the case of curvilinear steel bar structures, which are characterized by a variety of topologies, this is of great importance. In turn, the creation of a structural model using the Karamba 3D plug-in gave the possibility of an initial verification of the geometry, in relation to the structural requirements, which led to effective shaping.

5. Conclusions

The form of the roof received by algorithmic-aided shaping has been appropriately adapted to environmental conditions, whereas external conditions have caused the adaptability of the structural members of the roof. Moreover, the choice of the minimal surface as the base surface for the design of the roof form brings great benefits. First of all, the amount of cladding needed to make the roof could be reduced. In addition, the use of a minimal surface as the surface for the location of the structural nodes can influence the beneficial arrangement of forces in the structure. Thanks to this, it was possible to reduce the cross-sections of the bars, and hence the mass of the structure. Dimensioning was carried out for two bar cross-sections, which also gave the opportunity to choose the optimal cross section, due to the mass of the structure.

The research presented in the paper shows how using the tools for generative design, the process of shaping curvilinear steel bar structures, could be potentially improved in order to create responsive structural forms that meet both architectural and structural requirements.

Due to the fact that, in the case of bar structures, the unification of structural members is very important, the optimization of this issue will be the subject of further considerations of the author.

Funding: This research was funded by Rzeszow University of Technology.

Conflicts of Interest: The author declares no conflict of interest.

References

1. Vitruvius, P. *The Then Books on Architecture*, 1st ed.; Morgan, M.H., Ed.; Harvard University Press: Cambridge, UK, 1914; pp. 13–31.
2. Biermann, V.; Borngasser, B.; Evers, B.; Freigang, C.; Gronert, A.; Jobst, C.; Kremeier, J.; Lupfer, G.; Paul, J.; Ruhl, C.; et al. *Architectural Theory from the Renaissance to the Present*, 1st ed.; Taschen: Koln, Germany, 2003; pp. 6–20, ISBN 978-3-8365-5746-7.
3. Woliński, S. On the criteria of shaping structures. *Sci. Papers Rzeszow Univ. Technol.* **2011**, *276*, 399–408.
4. Kuś, S. General principles of shaping the structure. In *General Building Construction, Building Elements and Design Basics*; Arkady: Warsaw, Poland, 2011; Volume 3, pp. 12–71. (In Polish)
5. PN-EN 1990:2004. *Eurocode. Basis of Structural Design*; PKN: Warsaw, Poland, 2004. (In Polish)
6. PN-EN 1991-1-1:2004. *Eurocode 1. Actions on Structures. Part 1-1: General Actions—Densities, Self-Weight, Imposed Loads for Buildings*; PKN: Warsaw, Poland, 2004. (In Polish)
7. PN-EN 1993-1-1:2006. *Eurocode 3. Design of Steel Structures. Part 1-1: General Rules and Rules for Buildings*; PKN: Warsaw, Poland, 2006. (In Polish)
8. PN-EN 1993-1-10:2005. *Eurocode 3: Design of Steel Structures—Part 1-10: Material Toughness and Through-Thickness Properties*; PKN: Warsaw, Poland, 2005. (In Polish)
9. Dzwierzynska, J.; Prokopska, A. Pre-Rationalized Parametric Designing of Roof Shells Formed by Repetitive Modules of Catalan Surfaces. *Symmetry* **2018**, *10*, 105. [CrossRef]
10. Elango, M.; Devadas, M.D. Multi-Criteria Analysis of the Design Decisions in architectural Design Process during the Pre-Design Stage. *Int. J. Eng. Technol.* **2014**, *6*, 1033–1046.
11. Luo, Y.; Dias, J.M. Development of a Cooperative Integration System for AEC Design. In *Cooperative Design, Visualization, and Engineering, CDVE 2004*; Lecture Notes in Computer Science; Luo, Y., Ed.; Springer: Berlin/Heidelberg, Germany, 2004; Volume 3190.
12. Wang, J.; Chong, H.-Y.; Shou, W.; Wang, X.; Guo, J. BIM—Enabled design Collaboration for Complex Building. In Proceedings of the International Conference on Cooperative Design, Visualization and Engineering (CDVE2013), Mallorca, Spain, 22–25 September 2013; Springer: Berlin/Heidelberg, Germany, 2013.
13. Kim, D.-Y.; Lee, S.; Kim, S.-A. Interactive decision making Environment for the Design optimization of climate Adaptive building Shells. In Proceedings of the International Conference on Cooperative Design, Visualization and Engineering (CDVE 2013), Mallorca, Spain, 22–25 September 2013; Springer: Berlin/Heidelberg, Germany, 2013.
14. Oxman, R. Theory and design in the first digital age. *Des. Stud.* **2006**, *27*, 229–265. [CrossRef]
15. Oxman, R. Thinking difference: Theories and models of parametric design thinking. *Des. Stud.* **2017**, *52*, 4–39. [CrossRef]
16. Dzwierzynska, J. Descriptive and Computer Aided Drawing Perspective on an Unfolded Polyhedral Projection Surface. *IOP Conf. Ser.* **2017**, *245*, 062001. [CrossRef]
17. Dzwierzynska, J. Single-image-based Modelling Architecture from a Historical Photograph. *IOP Conf. Ser.* **2017**, *245*, 062015. [CrossRef]
18. Dzwierzynska, J. Reconstructing Architectural Environment from a Panoramic Image. *IOP Conf. Ser.* **2016**, *44*, 042028. [CrossRef]
19. Dzwierzynska, J. Direct construction of Inverse Panorama from a Moving View Point. *Procedia Eng.* **2016**, *161*, 1608–1614. [CrossRef]
20. Biagini, C.; Donato, V. Behind the complexity of a folded paper. In Proceedings of the Conference: Mo. Di. Phy. Modelling from Digital to Physical, Milano, Italy, 11–12 November 2013; pp. 160–169.
21. Pottman, H.; Asperl, A.; Hofer, M.; Kilian, A. *Architectural Geometry*, 1st ed.; Bentley Institute Press: Exton, PA, USA, 2007; pp. 35–194, ISBN 978-1-934493-04-5.
22. Kolarevic, B. *Architecture in the Digital Age: Design and Manufacturing*, 1st ed.; Spon Press: London, UK, 2003; pp. 20–98, ISBN 0-415-27820-1.
23. Barlish, K.; Sullivan, K. How to measure the benefits of BIM—A case study approach. *Autom. Constr.* **2012**, *24*, 149–159. [CrossRef]
24. Wortmann, T.; Tuncer, B. Differentiating parametric design: Digital Workflows in Contemporary Architecture and Construction. *Des. Stud.* **2017**, *52*, 173–197. [CrossRef]
25. Hardling, J.E. Meta-parametric Design. *Des. Stud.* **2017**, *52*, 73–95. [CrossRef]

26. Stravic, M.; Marina, O. Parametric Modeling for Advanced Architecture. *Int. J. Appl. Math. Inform.* **2011**, *5*, 9–16.

27. Bhooshan, S. Parametric design thinking: A case study of practice-embedded architectural research. *Des. Stud.* **2017**, *52*, 115–143. [CrossRef]

28. Turrin, M.; von Buelow, P.; Stouffs, R. Design explorations of performance driven geometry in architectural design using parametric modeling and genetic algorithms. *Adv. Eng. Inform.* **2011**, *25*, 656–675. [CrossRef]

29. Dźwierzynska, J. Shaping curved steel rod structures. *Tech. Trans. Civ. Eng.* **2018**, *8*, 87–98. [CrossRef]

30. Autodesk. Available online: https://www.autodesk.com/ (accessed on 21 February 2019).

31. Preisinger, C. Linking Structure and Parametric Geometry. *Arch. Des.* **2013**, *83*, 110–113. [CrossRef]

32. Veenendaal, D.; Block, P. An overview and comparison of structural form finding methods for general networks. *Int. J. Solids Struct.* **2012**, *49*, 3741–3753. [CrossRef]

33. Dzwierzynska, J. Computer-Aided Panoramic Images Enriched by Shadow Construction on a Prism and Pyramid Polyhedral Surface. *Symmetry* **2017**, *9*, 214. [CrossRef]

34. PN-EN 1991-1-1:2004. *Eurocode 1. Actions on Structures. Part 1-3: General Actions—Snow Loads*; PKN: Warsaw, Poland, 2004. (In Polish)

buildings

MDPI

Article

Digitally Designed Airport Terminal Using Wind Performance Analysis

Lenka Kabošová [1,*], Stanislav Kmeť [2] and Dušan Katunský [1]

[1] Institute of Architectural Engineering, Faculty of Civil Engineering, Technical University of Kosice, 042 00 Kosice, Slovakia; dusan.katunsky@tuke.sk

[2] Institute of Structural Engineering, Faculty of Civil Engineering, Technical University of Kosice, 042 00 Kosice, Slovakia; stanislav.kmet@tuke.sk

* Correspondence: lenka.kabosova@tuke.sk; Tel.: +421-055-602-4157

Received: 17 January 2019; Accepted: 11 February 2019; Published: 7 March 2019

Abstract: Over the past few decades, digital tools have become indispensable in the field of architecture. The complex design tasks that make up architectural design methods benefit from utilizing advanced simulation software and, consequently, design solutions have become more nature-adapted and site-specific. Computer simulations and performance-oriented design enable us to address global challenges, such as climate change, in the preliminary conceptual design phase. In this paper, an innovative architectural design method is introduced. This method consists of the following: (1) an analysis of the local microclimate, specifically the wind situation; (2) the parametric shape generation of the airport terminal incorporating wind as a form-finding factor; (3) Computational Fluid Dynamics (CFD) analysis; and (4) wind-performance studies of various shapes and designs. A combination of programs, such as Rhinoceros (Rhino), and open-source plug-ins, such as Grasshopper and Swift, along with the post-processing software Paraview, are utilized for the wind-performance evaluation of a case study airport terminal in Reykjavik, Iceland. The objective of this wind-performance evaluation is to enhance the local wind situation and, by employing the proposed architectural shape, to regulate the wind pattern to find the optimal wind flow around the designed building. By utilizing the aforementioned software, or other open-source software, the proposed method can be easily integrated into regular architectural practice.

Keywords: performance-oriented design; parametric architecture; form-finding; CFD; wind analysis; digital tools

1. Introduction

The development of digital tools has created new options for future architectural design, especially in regards to digital tools that operate as open-source. Therefore, the designers take an interest in exploring such tools for a variety of design tasks. "The emerging open-source attitude over proprietary information may lead to "the true" digital revolution" [1]. By using a digital, computer-based approach to design, factors, such as specific weather conditions, or topographic conditions of the individual design sites, can be incorporated into the form-finding process early in the design phase. All of the above can complement the vision of the architect, with the result being nature-conscious architecture. From being merely a "statue" inserted into a given location, these design tools can help architecture to become a more complex mechanism that originates from, and is bound up with, its surroundings.

1.1. Natural Environment

As climate change continues to create shifting weather patterns, it is of great importance that architecture becomes an integral part of its surrounding environment. The latest assessment by the Intergovernmental Panel on Climate Change (IPCC) [2] warned that global warming is unequivocal and is caused most certainly by human activity. As a consequence, not only is the world uniting in the political sphere to find solutions for cities to resist climate change [3], but also architecture has begun to center on creating a reciprocal relationship with nature. Innovative design strategies that focus on incorporating climatic circumstances into the design process are resulting in environmentally-responsive architecture. In the ever-changing environment, it seems to be a fair assumption that adapting building design to climate change is unavoidable. However, questions regarding the extent of the change required and how we can achieve such change remain unanswered. By acting proactively, we can make adjustments to current design strategies that can help to mitigate the worst-case scenarios for the future environment [4–7]. Temperature rises are the most evident effect of global warming, demanding a reaction from the architectural field. However, wind storms and sudden wind gusts are much harder to predict, due to their chaotic nature. A flexible architectural response is required in various architectural disciplines, including urbanism, architecture, or dynamic structures.

1.2. Wind-Formed Architecture

Wind is a fluctuating element of nature, its speed and direction can change quickly and, as such, it can have a very significant impact on both the natural and artificial environment. The phenomenon of wind erosion contributes to shaping the natural environment. Similarly, the built (artificial) environment is affected by wind flow and reciprocally, by the relative position and shape of buildings, alters wind flow. By taking this premise into account early in the conceptual design stage, improvements or modifications can be made to the local wind situation. Regarding architectural obstacles to the wind, the flow of the wind can be deflected, concentrated, diffused, harvested, or minimally changed [8]. Aerodynamically designed buildings, for instance, contribute to reducing deflections of the wind and the resonance caused by the wind. On an urban scale, wind-related planning can help to avoid the occurrence of wind tunnel effects, as well as lower the air pressure on buildings [6]. By optimizing the shape and dimensions of buildings, natural ventilation can be enhanced [9]. However, the negative effects of wind can be a concern for planners for several other reasons. Dry wind carries sand and dust particles, which can cause damage to building surfaces and clog filters and ventilation ducts. Moreover, humid sea wind in coastal areas contains sodium chloride, which can cause corrosion [5].

1.3. Performative Architecture

For an environmentally-adapted design, the wind and/or lighting criteria are an inseparable part of the form-finding process. The interconnection of wind flow and architecture, specifically internal and external airflow simulations, are receiving increased prominence in architectural design [10]. Based on the wind or solar simulation outcomes, the performance of the design can be continuously improved until the most suitable solution is chosen.

"Performance-based design is an approach that leverages iterative simulation as a way of exploring design options" [10]. Digital tools provide an opportunity to support performance-based conceptual design and enable researchers to create complex simulations, such as computational fluid dynamics (CFD) and wind analysis, and substitute a physical wind tunnel. Wind-related architectural performance-based design is usually carried out in the final design stage with very few possible alterations to the actual design. Nevertheless, it is necessary to adopt this approach from the initial phase of planning if the goal is to improve the environmental performance of the design [11]. Additionally, apart from high-rise buildings, it is not general practice to analyze other types of buildings.

Furthermore, employing digital tools in the design process enables researchers to conduct parametric studies, and to generate multiple design options. This allows for the evaluation of several design options with different sets of parameter variables. However, as the number of possible alternative solutions grows, it becomes infeasible for the designer to assess every proposed solution, which might be perceived as a drawback [12,13]. One solution for this could be the optimization of buildings by using genetic algorithms as a form-finding method. This type of method might lead to an improved design and a way to entirely avoid subjective design decisions [13,14]. However, in most cases the planner can reasonably determine which of the various parametrically-generated design options (form generation based on geometric constraints) [15] suits the design situation the best and can then choose it for further work.

1.4. CFD (Computational Fluid Dynamics) Analysis in Architecture

If the wind is the main driving factor in the parametric form-finding process, the desired interaction of architecture and the wind determines how the parameters are set to achieve a wind-responsive final design. Simultaneously, the wind performance of the various design alternatives is investigated. Several examined design alternatives created using parametric software, such as the visual programming language *Grasshopper* for *Rhino* and consequently tested using CFD can lead to the improved building shape or urban layout. On the one hand, such wind analysis in the early design stage, coupled with parametric designing is not yet smoothly integrated into the design loop [16]. On the other hand, "many projects have demonstrated the usefulness of parametric studies with CFD simulations both for identifying the key parameters in the simulation, as well as for exporting different design alternatives", as Malkawi et al. remarked already in 2005 [17].

CFD has rapidly improved over the last 30 years [18]; it has become gradually accessible to design practices [19], and is slowly penetrating into the architectural and civil engineering area [20]. The fact that the positive and negative space of buildings in an urban layout shapes air movement was addressed in 1998 [21]. Consequently, the application of CFD simulations on building structures exhibits great potential for not only understanding, but also improving the dynamic interactions of the wind with the built environment [22].

The wind flow's potential to generate the shapes of building envelopes is examined in the following case study. This design strategy investigates the parametric shape generation of an airport terminal and subsequently examines the wind performance of three shape alternatives using open-source plug-ins for *Rhino*. The interaction of the proposed shapes with the wind will be observed, alongside their effect on the pedestrian wind comfort near the intended entrance.

2. Materials and Methods

The methodology presented in this research is utilized to find the best performing building envelope in the specific wind conditions. At the same time, however, this step-by-step architectural design method proposing a design loop for creating building envelopes is the intention of this study. Figure 1 is a scheme of the developed methodology incorporating the detailed analysis of the wind situation through parametrically creating various shape alternatives, and subsequent CFD testing to finally selecting the best-performing geometry, decided by the designer according to the initial goals.

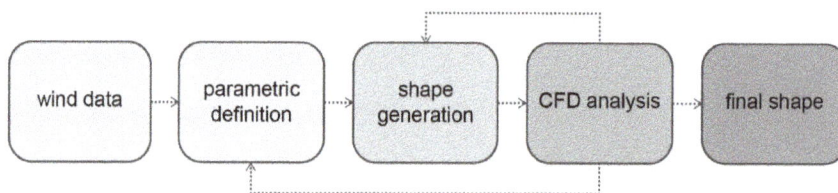

Figure 1. A scheme of the developed design loop. CFD: Computational Fluid Dynamics.

For an illustration of the design loop that enables us to search for the optimal wind-influenced architectural shape in the iterative process, the airport in Reykjavik is selected as a case study site. The airport location is selected for the following reasons: (1) Iceland is a remote island very close to the North Polar Circle (Arctic Circle), which influences its climatic conditions; (2) the airport has interesting micro-climatic conditions caused by the high hill on the east and the bay on the south-west; (3) the old airport building, shaped as a simple block with a pitched roof is placed very close to the short runways [23] and creates an obstacle to the wind. Instead, an aerodynamically-shaped alternative is sought within the framework of this case study.

2.1. Wind Speed in the Last Decades

Firstly, we want to understand the climatic circumstances, particularly the wind situation in Iceland and the micro-climatic conditions in Reykjavik. Scientists use several types of climatic models for climatic data evaluation. Climate Reanalyzer is a free weather map web database, created by the University of Maine [24]. In this case study, the maps are used to find the correlation of two time periods of 1999-2016, and 1979-1998. The data are exported as images using two different climatic models, ERA-Interim and NCEP/NCAR Reanalysis VI (Figures 2 and 3). The figures show the tendency of the wind speed at a 10 m height above the ground. Two selected climatic models interpret the changes with slight differences. Both of them, however, indicate higher wind speeds in the last years compared to the period from 1979 to 1998, at least over a part of Iceland. In this area, the strong wind is the most frequent weather phenomenon that often impacts airport operations [23]. Based on the known facts, we can assume a tendency of accelerating wind in this location. Nevertheless, the wind behavior is fluctuating and complex, and the predictions for the future development of the wind speed are even more unclear.

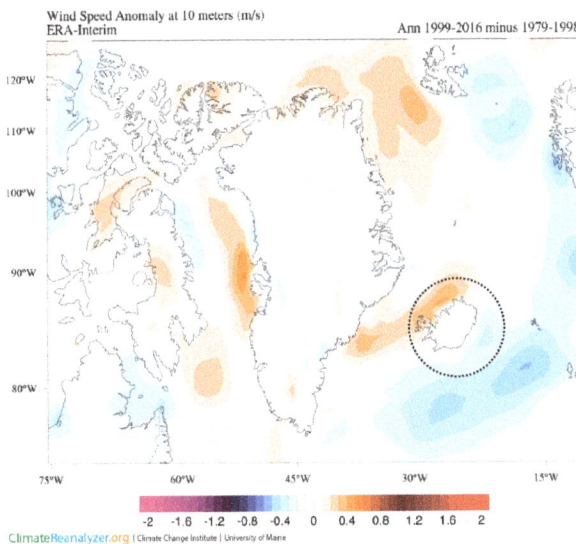

Figure 2. The ERA-Interim climatic model. The red color indicates a higher annual mean wind speed in recent years (1999-2016), while the blue color shows the opposite. Dataset source: https://apps.ecmwf.int/datasets/.

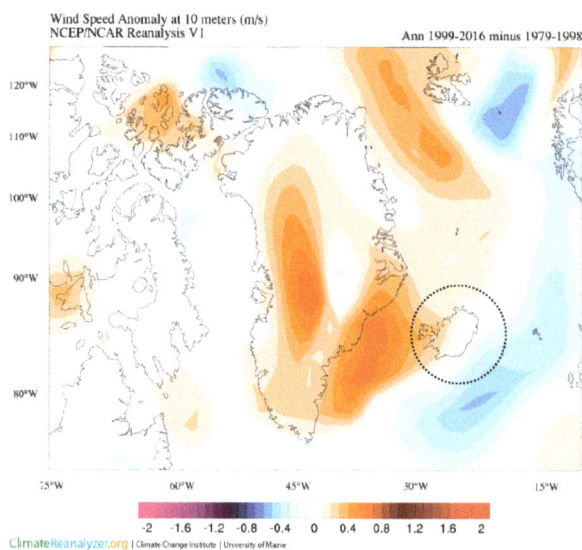

Figure 3. The NCEP/NCAP Reanalysis VI climatic model. The red color indicates a higher annual mean wind speed in recent years (1999-2016), while the blue color shows the opposite. Dataset source: http://www.cpc.ncep.noaa.gov/products/wesley/reanalysis.html.

2.2. Weather Data Statistics

This project works only with the digitally-accessible data, as this first stage of the project is conducted in a digital, computer-based way. The weather data for Reykjavik were downloaded as an *epw weather file from the EnergyPlus online weather database [25]. The file contains representative data for the chosen location.

3. Reykjavik Airport Analysis

Reykjavik (RKV) airport is the second largest airport in Iceland and is only 50 kilometers away from Keflavik (KEF) airport. The location for the city's international airport was selected in 1940, and the small airport terminal was constructed in 1948. In the 1960s, the two main flight companies moved their operations to Keflavik, while keeping domestic flights in Reykjavik. The Reykjavik airport is slated for closure in 2024, with the main reason being the possible development of the area, which is close to the city center. In any case, no final decision has yet been taken.

The airport has three short runways; two of them are active, but the shortest (869 meters long) runway, was closed in 2016 (Figure 4). The other two runways are 01/19, which is 1567 m long, and 13/31, which is 1230 m long. Surrounded by the bay from the south-west side and a 60-meter high mountain from the east side, the airport has, naturally, a flattened morphology. Family houses are located north of the airport [23]. The airport in Reykjavik had 376 347 passengers last year, which was 5.5% less than the year before that [26].

The airport site serves as a case study site. The placement of the new terminal building is considered at the same location as the old terminal, close to runway 01/19.

Figure 4. The airport in Reykjavik. The satellite map is retrieved from Google Earth.

Wind Rose

As mentioned above, EnergyPlus is used to obtain the *epw weather file that is subsequently imported to *Grasshopper*. *Ladybug* for *Grasshopper* is a collection of free environmental design applications that can convert a weather file into a graphical form of wind rose. The wind speed and direction can be displayed for a specified period throughout the year, while also showing the frequency of winds (Figure 5). The analysis indicates that easterly winds are prevailing (15.64% of the time per year) at the airport in Reykjavik. The wind gust speed can reach 25 m/s and more in all wind directions. Barely 0.72% of the time, which accounts for 63 hours per year, there is no wind in the given area. The average wind speed for easterly winds is 6.2 m/s. This value can be obtained from the weather file and is used for wind analysis in the following chapters. According to a different data source available on the Icelandic Meteorological Office's website based on the measured values in the years 1949-2017, the average wind speed for all wind directions in Reykjavik is 5.53 m/s [27].

According to data from the EnergyPlus website, the average wind speed for all wind directions is around 5.89 m/s, which is only a slightly different value. This data source is hence chosen, due to its convenient usage in *Grasshopper*. The prevailing easterly winds are stronger than 5 m/s for 9.38% of hours per year. For all wind directions, the wind speed is higher than 5 m/s for 52.4% of hours per year. Moreover, the wind speed exceeds 15 m/s for 2.59% of hours per year, with the south-east direction encountering the strongest wind gusts. According to the Dutch Standard (NEN 8100: Wind comfort and wind danger in the built environment) that sets the threshold values for pedestrian wind comfort, if the value of 5 m/s is exceeded more than 20% of the time, the conditions for outdoor pedestrian activities are poor. If the wind speed is higher than 15 m/s for 0.3% hours per year, the conditions become dangerous [28]. The average wind temperature in Reykjavik is relatively low which contributes to the pedestrian discomfort in the outdoor environment (Figure 6). Only for 13.57% of hours per year is the temperature of the winds higher than 10 °C.

These observations are taken into account in the new airport building design.

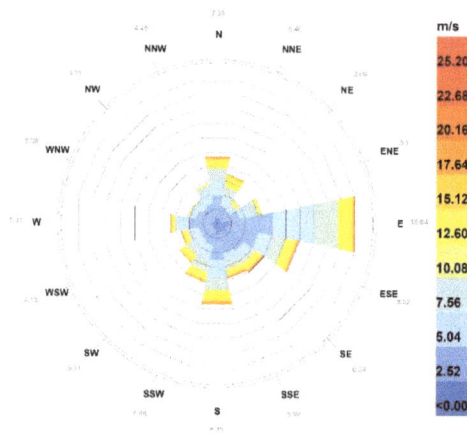

Figure 5. The windrose of Reykjavik's airport, Iceland. Displayed are hourly wind speed data in m/s for the period from January 1 to December 31. The location is calm (wind speed = 0 m/s) 0.72% of the time.

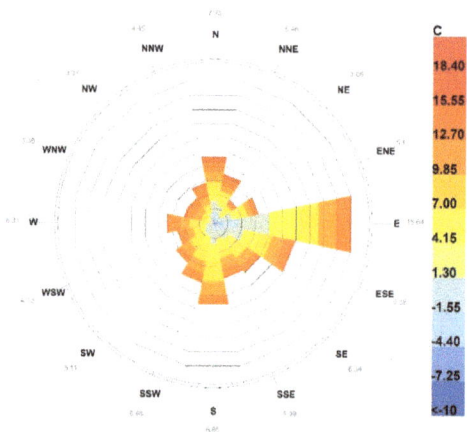

Figure 6. The windrose of Reykjavik's airport, Iceland. Displayed is the hourly temperature in °C for the period from January 1 to December 31.

4. Parametrically-Controlled Design

Parameter-based algorithmic modeling offers greater freedom for designers in the exploration of possible solutions and the creation of best-fitting alternatives [1]. Determining the geometry-generating process, which of the parameters can change, and how they influence the final design is key to environment-reactive solutions.

In the given microclimate of Reykjavik's airport, based upon the idea of the interaction with the wind, a wind-shaped airport terminal is designed. It is intended as a conjunction of the finger/pier and the linear airport terminal concept [29], but the building program is not elaborated further, as it is not relevant for this phase of the research. The Reykjavik's airport terminal currently in service is a two-floor rectangular building, with a floor plan size of around 80 × 50 meters. A possible slight increase in the number of passengers is accounted for in the design. Regarding the relation to the wind, the new terminal is designed as a streamlined body. The dimensions differ a little for every shape modification (Table 1). An optimal shape is sought to interact with the wind with minimum resistance to the flow, providing a pleasant outdoor climate around the entrance to the airport terminal, as well as the outdoor wind shelters, while not influencing airplanes' take-offs and landings on the runways in the proximity of the terminal.

Table 1. Combinations of parameters set in *Grasshopper* for three variants.

SET nr.	Preview	Floor-Pland Imensions	Number of Shape-Influencing Points	Extrusion Height
1		130 × 57 m	3	10
2		125 × 65 m	3	12.5
3		120 × 72 m	3	11.5

Airport Terminal

The parametric control of the design and the overall shape generation is performed by utilizing *Grasshopper*. The airport terminal's shape is created from a single, closed planar curve. The shape formation is also dependent on 'shape-influencing' points, behaving as attractors. These core elements are variable and are designed and set at the beginning of the wind-influenced form-finding process, based on the knowledge of the site's wind situation (Figure 7). The proposed parametric definition consists of the following main steps:

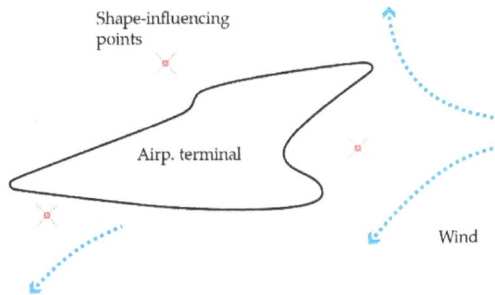

Figure 7. The core elements for the shape generation. The schematic wind flow is displayed in blue color.

(1) The designed planar curve is defined by the constant number of control points. In this research, the number is set to 25.

(2) The proximity (the distance) of curve control points to the 'shape-influencing' points located on the same plane is determined.

(3) Based on the ascertained distances, each control point is lifted in the z-direction in the pre-specified height range.

(4) A double-curvature shape of the morphing roof/façade is created by moving the 'shape-influencing' attractor points that represent the shaping power of the wind flow and the vision of the designer at the same time. The designer can set any number of shape-influencing points for integration into the form-finding process. In this research, there are three points used. The closer the points are to the curve, the more the shape is closed to the wind flow, and ergo the curve control points

are kept at zero height. The curve control points can move only in the z-direction and only to the maximum height, restricted to 20 meters.

(5) In the next step, the volume is created consisting of slices of various extrusion heights, set by the designer (Table 1).

After the preliminary tests of several diversified base curves, only one type is selected for the further elaboration. Three shape modifications of the previously selected closed planar curve with different positions of the shape-influencing points (three parameters sets) generate the three airport terminal shapes tested in this paper.

The generated shapes are displayed in Figures 8–10. The sliced, rough geometry is intended to decelerate the wind flow. The geometry of the airport terminal is at some points lifted above the ground and at some points touches the ground to create a sheltered open-air space.

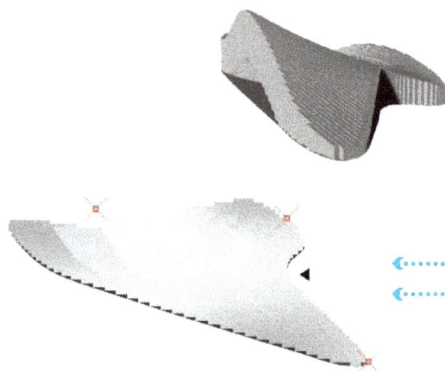

Figure 8. The first set of variable parameters. The floor plan depicting shape-influencing points (red) and wind direction (blue arrows). The entrance is depicted as a black arrow. Above is the 3D shape of the intended airport terminal.

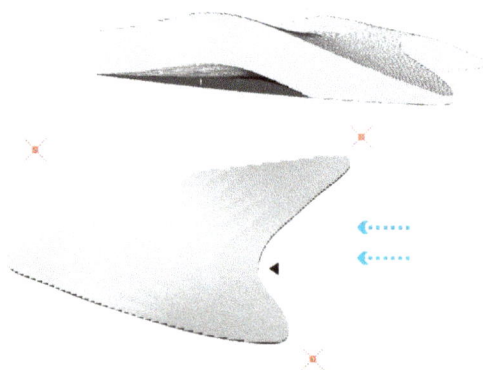

Figure 9. The second set of variable parameters. The floor plan depicting shape-influencing points (red) and wind direction (blue arrows).

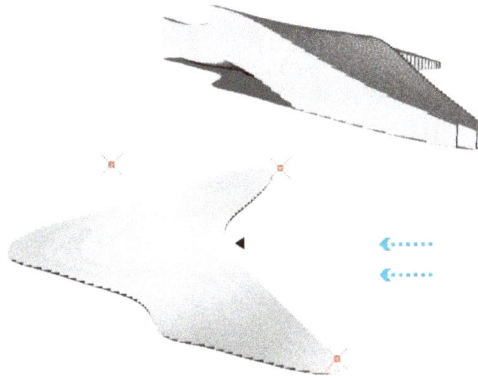

Figure 10. The third set of variable parameters. The floor plan depicting shape-influencing points (red) and wind direction (blue arrows).

5. CFD simulations in Swift

Swift CFD software is used to perform the wind tests of the three parametrically-controlled shape alternatives. The simulation results are compared in *Paraview*, an external open-source software necessary for visualizing the CFD results. Despite the wind speed and direction fluctuating in time, there is a clear prevailing easterly wind direction at Reykjavik's airport. This wind direction and the average wind velocity of 6.2 m/s are used in the simulations.

5.1. Swift Software

ODS Engineering has developed the *Swift* CFD tool for the working environment of *Grasshopper*. It runs on the platform of *OpenFOAM* (open source CFD toolbox). Originally, *ODS Studio*, combining powerful engines, such as *OpenFOAM*, *Radiance*, and *EnergyPlus* into one simulation tool, was developed for *Blender*. *Swift* is a new plug-in for *Grasshopper* that functions on an Ubuntu virtual machine running on Windows 10. The process of installation requires following the YouTube tutorial provided by ODS Engineering. A big advantage of this new CFD software integrated with *Grasshopper* is that the whole design process, starting from an architectural idea, through the parametric control of the model and subsequent CFD simulation of every modification, can be done in one environment. Another benefit is the programming approach to the setup of CFD cases based on the *OpenFOAM* syntax rules. *Swift*, however, requires *Paraview* external software for the visualization of the calculated results. Also, there is almost no online support, or a guide for the beta version of *Swift*, which remains a disadvantage at this moment, although a user can study instructions for *OpenFOAM* in general and make use of them in *Swift*.

For this research, the 'SimpleFoam' (semi-implicit method for pressure-linked equations) algorithm is used as an iterative algorithm for solving the Navier-Stokes equations, whereas RAS (Reynolds-averaged stress) is used as a turbulence model for a steady-state simulation [30]. Block Mesh and Snappy HexMesh utilities are used to mesh the examined geometry.

5.2. Input Settings and Boundary Conditions

The simple external CFD' algorithm, accessible online, is used. The following changes in the boundary conditions are made: (1) the inlet wind speed is set to 6.2 m/s at the maximum x; (2) the wind tunnel walls are set to 'wall slip'; and (3) after several trial attempts, the solver is set to terminate after 250 iterations when the residual of the equations reaches the convergence tolerance of 10^{-5} for the velocity components, Ux, Uy, and Uz [30].

The minimum level of geometry meshing is set to 4, whereas the wind tunnel cell size is set to 3 meters. According to the best CFD guides, the wind tunnel dimensions should be integral multiples of the tested geometry height. Upwind and downwind, the size of the domain is 5H, and 15H, respectively, while the sides, as well as the height, is 5H [31]. *Swift* settings cannot precisely control the geometry placement in the wind tunnel, which is why the CFD guidelines cannot be applied here. The geometry is centered in the *xy* plane. The results of the steady-state simulations are shown in Figures 11–13. For a better visual comparison of the results in *Paraview*, the wind velocity data range is set from 0 to 10 m/s for each case. The blue color indicates a wind speed close to zero, whereas the red color indicates the maximum wind speeds. The horizontal slice for projecting the results is made 2 meters above the ground to capture the wind situation at the pedestrian level. The 3D streamlines with wind vectors are added to get a better idea of the air movement around the shapes.

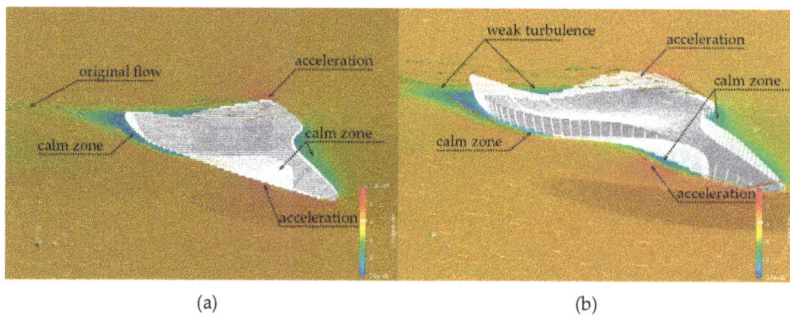

(a)   (b)

Figure 11. CFD results of shape variation 1: top view (**a**) and perspective view (**b**).

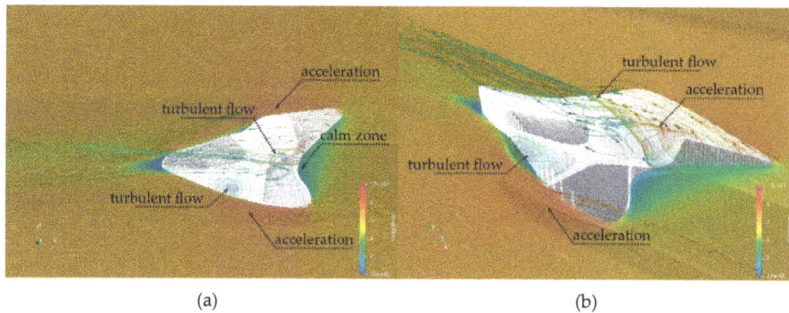

(a)   (b)

Figure 12. CFD results of shape variation 2: top view (**a**) and perspective view (**b**).

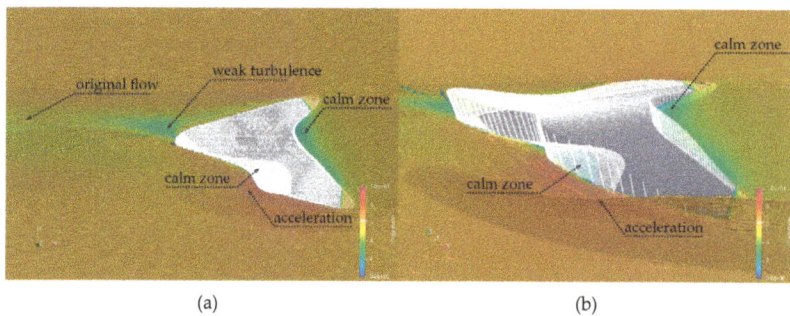

(a)   (b)

Figure 13. CFD results of shape variation 3: top view (**a**) and perspective view (**b**).

6. Discussion

Using the parametric approach, the change of selected parameters and the core geometry enables a fast change of the resulting 3D geometry. Subsequently, every new shape is referenced in the (again parametric) virtual wind tunnel, although the CFD settings and boundary conditions are kept the same for each option. The automatic meshing creates a different number of calculation cells for every option, which is around 4.4–5.4 million. Three sets of parameters create shape variants of the airport terminal building which interact with the wind flow differently. By observing and examining the CFD wind analysis results in *Paraview*, each shape option, being in principle a streamlined body, represents a minimum resistance to the wind flow. The shape variation number 3 is, as concluded from the obtained results, the most aerodynamic. Also, the intended sheltered open-air space facing the south is quite calm in variant 3, as well as in variant 1. The entrance area for all three shapes is relatively calm, with the wind speed ranging from 0 to around 4 m/s. The shape of variant 2 appears to be aerodynamic; however, the wind flow, influenced and transformed by the shape, turns into a turbulent flow.

6.1. Advantages of the Approach

(1) The parametric performative design enables us to verify more design options already in the conceptual design stage.

(2) The changes in the core geometry, as well as the changes in several parameters of the definition, result in the improvement of the design's environmental performance.

(3) The architecture designed for the specific wind conditions can be digitally tested, and the crucial design decisions can be made based on the computer simulations results very early in the design process.

(4) By utilizing digital tools in the design process, the otherwise time and money-consuming wind analyses are accessible to everyone and practically free of cost in the case of open-source software.

(5) The proposed combination of *Rhino*, *Grasshopper*, *Swift*, and *Paraview*, which are all, except for *Rhino*, open-source digital tools that can create, generate, and test the designed geometry in one working environment.

(6) The wind-formed parametric geometry is the first step in the architectural design process. The further elaboration of the project, including the structural design, as well as the building program, can be consequentially performed in a similar, parametric manner, integrated into the original parametric definition.

6.2. Disadvantages of the Approach

(1) Only a limited number of geometries can be tested as the necessity of manually setting the parameters, especially manually changing the core geometry, requires time. As was mentioned earlier, this drawback has a solution. Genetic algorithms automatically test hundreds of options. However, the goal of the approach is not to test every possible shape within the set of parameters, but to direct the designer's own decisions from the beginning of the design process.

(2) The results of the shape-generating or form-finding process depend on the initial parametric definition, which has to be adequately set.

(3) The approach using *Swift* for *Grasshopper* requires using *Paraview* as external software for the visualization of the results. This way, verifying more alternative solutions is more time-consuming.

(4) The interpretations of the CFD simulation results are made by the designer according to the observation of the wind speed and flow pattern around the designed architectural shape. To mathematically support the designer's opinion, additional probe points can be placed at the observed locations to measure the exact wind speed at the selected location.

7. Conclusions

Parametric designing, incorporating the wind phenomenon along with the actual wind microclimate of the site, is a step towards nature-responsive architecture. Ways of connecting the artificial and the natural environment are becoming even more sought-after with the progression of global climate change. The analogy of the shaping effects of the wind in nature can be, in a figurative sense, transformed into wind-shaped architecture. Architecture created from the original idea with the wind phenomenon as the driving force in the form-finding process would be a big benefit for extremely windy areas. The paper presents a case study in Reykjavik, Iceland, where strong, cold winds are the climate's characteristics. The digital open-source attitude to architectural design is a basis for the proposed design strategy. The use of the parametric approach in the design of the airport terminal enables the examination of various design options and their performance in wind flow. Three shape modifications of the closed planar curve, with various positions of the 'shape-influencing' points, create three building envelopes to demonstrate the use of the parametric approach in the wind-related design. The three aerodynamic shapes are analyzed in Reykjavik's easterly winds using a parametrically adjustable CFD virtual wind tunnel definition. The analysis of the three observed options directs the design process towards an architectural solution that is adapted to the specific wind conditions. Including the digital CFD analysis into the architectural design in the early conceptual phase contributes to defining and predicting the effects of architecture on the wind fluxes in its vicinity. For the correct interpretation of the wind simulation results, a close observation of the wind behavior is required. *Swift* enables the use of points which act as probes for measuring the wind speed at the locations of interest, in case there is a need to verify the interpreted results.

The CFD results of the three tested shapes indicate that the most suitable shape for the next design steps is variation number 3. The shape performs the best according to the following requirements: (1) it has a calm entrance area, (2) a calm outdoor wind shelter and (3) minimum resistance to the wind flow, ergo minimum turbulence around the airport terminal.

The form-finding process of wind-shaped architecture presented in the paper will be further elaborated in the following research. The most suitable building's structure will be parametrically sought after.

Author Contributions: Conceptualization, L.K., D.K. and S.K.; methodology, L.K., D.K. and S.K.; software, L.K.; validation, L.K., D.K. and S.K.; formal analysis, L.K.; investigation, L.K.; resources, L.K.; data curation, L.K.; writing—original draft preparation, L.K.; writing—review and editing, L.K., D.K. and S.K.; visualization, L.K.; supervision, D.K. and S.K.; project administration, L.K., D.K. and S.K.; funding acquisition, D.K. and S.K.

Funding: This research was funded by VEGA 1/0302/16and VEGA 1/0674/18 (Grant Agency of the Slovak Republic).

Acknowledgments: Open-source software used in the project: *Grasshopper*: https://www.grasshopper3d.com/; *Ladybug* for *Grasshopper*: https://www.grasshopper3d.com/group/ladybug; *Swift* for *Grasshopper*: https://www.ods-engineering.com/tools/ods-swift/; *Paraview*: https://www.paraview.org/.

Conflicts of Interest: The authors declare no conflict of interest. The funders had no role in the design of the study; in the collection, analyses, or interpretation of data; in the writing of the manuscript, or in the decision to publish the results.

References

1. Barlieb, C.; Richter, C.; Greschner, B.; Tamke, M. Whispering Wind: Digital Practice and the Sustainable Agenda. In *Computation: The New Realm of Architectural Design, Proceedings of the 27th eCAADe Conference, Istanbul, Turkey, 2009*; Çağdaş, G., Colakoglu, B., Eds.; Istanbul Technical University: Sarıyer, Turkey, 2009; pp. 543–550.

2. IPCC. *Climate Change 2014: Synthesis Report. Contribution of Working Groups I, II and III to the Fifth Assessment Report of the Intergovernmental Panel on Climate Change*; Core Writing Team, Pachauri, R.K., Meyer, L.A., Eds.; IPCC: Geneva, Switzerland, 2014; 151p, ISBN 978-92-9169-143-2.

3. Bassolino, E.; Ambrosini, L. Parametric environmental climate-adaptive design: The role of data design to control urban regeneration project of BorgoAntignano, Naples. *Proc. Soc. Behv. Sci.* **2015**, *216*, 948–959. [CrossRef]
4. Kuenstle, M.W. Flow structure environment simulation: A comparative analysis of wind flow phenomena and building structure interaction. In *Connecting the Real and the Virtual: Design e-Ducation, Proceedings of the 20th eCAADe Conference, Warsaw, Poland, 2002*; Koszewski, K., Wrona, S., Eds.; Warsaw University of Technology: Warsaw, Poland, 2002; pp. 564–568.
5. Kuismanen, K. Climate—Conscious Architecture: Design and Wind Testing Method for Climates in Change. Ph.D. Thesis, University of Oulu, Oulu, Finland, 28 November 2008.
6. Snow, M.; Prasad, D. Climate Change Adaptation for Building Designers: An Introduction, Environment Design Guide 66 MSa. 2011. Available online: Environmentdesignguide.com.au (accessed on 23 October 2018).
7. Kerestes, J.F. Design out of necessity: Architectural approach to extreme climatic conditions. In *Design in Freedom, Proceedings of the 18th SIGRADI Conference, Montevideo, Uruguay, 2014*; Blucher: Sao Paulo, Brazil, 2014; pp. 130–133.
8. Kormaníková, L.; Achten, H.; Kopřiva, M.; Kmeť, S. Parametric wind design. *Front. Archit. Res.* **2018**, *7*, 383–394. [CrossRef]
9. Muhsin, F.; Yusoff, W.F.M.; Mohamed, M.F.; Sapian, A.R. The Effects of Void on Natural Ventilation Performance in Multi-Storey Housing. *Buildings* **2016**, *6*, 35. [CrossRef]
10. Kaushik, V.; Janssen, P. Urban Windflow: Investigating the use of animation software for simulating windflow around buildings. In *Real Time: Extending the Reach of Computation, Proceedings of the 33rd eCAADe Conference, Vienna, Austria, 2015*; Martens, B., Wurzer, G., Grasl, T., Lorenz, W.E., Schaffranek, R., Eds.; Vienna University of Technology: Vienna, Austria, 2015; pp. 225–234.
11. Kim, H.-J.; Kim, J.-S. Design Methodology for Street-Oriented Block Housing Considering Daylight and Natural Ventilation. *Sustainability* **2018**, *10*, 3154. [CrossRef]
12. Erhan, H.; Wang, I.; Shireen, N. Interacting with thousands: A parametric-space exploration method in generative design. In *Design Agency, Proceedings of 34th ACADIA Conference, Los Angeles, California, 2014*; Huang, A., Sanchez, J., Gerber, D., Eds.; ACADIA/Riverside Architectural Press: Toronto, ON, Canada, 2015; pp. 619–625.
13. Wang, L.; Tan, Z.; Ji, G. Toward the wind-related building performative design: A wind-related building performance optimization design system integrating Fluent and Rhinoceros based on iSIGHT. In *Living Systems and Micro-Utopias: Towards Continuous Designing, Proceedings of the 21st CAADRIA Conference, Melbourne, Australia, 2016*; The University of Melbourne: Melbourne, Australia, 2016; pp. 209–218.
14. Mooneghi, M.A.; Kargarmoakhar, R. Aerodynamic Mitigation and Shape Optimization of Buildings: Review. *J. Build. Eng.* **2016**, *6*, 225–235. [CrossRef]
15. Grobman, Y.J.; Yezioro, A.; Capeluto, I.G. Computer-based form generation in architectural design—A critical review. *Int. J. Archit. Comput.* **2009**, *7*, 535–553. [CrossRef]
16. Chronis, A.; Dubor, A.; Cabay, E.; Roudsari, M.S. Integration of CFD in computational design. In *ShoCK!, Proceedings of the 35th eCAADe Conference, Roma, Italy, 2017*; Fioravanti, A., Cursi, S., Elahmar, S., Loffreda, S.G.G., Novembri, G., Trento, A., Eds.; Sapienza University of Rome: Rome, Italy, 2017; pp. 601–610.
17. Malkawi, A.M.; Srinivasan, R.S.; Yi, Y.K.; Choudhary, R. Decision support and design evolution: Integrating genetic algorithms, CFD and visualization. *Autom. Constr.* **2005**, *14*, 33–44. [CrossRef]
18. Cóstola, D.; Blocken, B.; Hensen, J.L.M. Overview of pressure coefficient data in building energy simulation and airflow network programs. *Build. Environ.* **2009**, *44*, 2027–2036. [CrossRef]
19. Alexander, D.K.; Jenkins, H.G.; Jones, P.J. A comparison of wind tunnel and CFD methods applied to natural ventilation design. In Proceedings of the Building Simulation IBPSA Conference, Prague, Czech Republic, 8–10 September 1997; pp. 1–7.
20. Moukalled, F.; Mangani, L.; Darwish, M. *The Finite Volume Method in Computational Fluid Dynamics: An Advanced Introduction with OpenFOAM and Matlab*; Springer International Publishing: Basel, Switzerland, 2016.
21. Tsou, J.Y. Applying computational fluid dynamics to architectural design development. In Proceedings of the 3rd Conference on Computer Aided Architectural Design Research in Asia, CAADRIA, Osaka, Japan, 22–24 April 1998; pp. 133–142.

22. Kuenstle, M.W. Computational Flow Dynamic Applications in Wind Engineering for the Design of Building Structures in Wind Hazard Prone Urban Areas. In Proceedings of the 5th SIGradi Conference, Concepcion, Chile, 21–23 November 2001; pp. 67–70.

23. Donohue, K. Saga of Two Icelandic Airports. *Airways*. 2016, pp. 37–42. Available online: http://www.kendonohue.com/articles/AW249_AirportReview.pdf (accessed on 16 December 2018).

24. CCI—Reanalyzer. Climate Change Institute, University of Maine, USA: Monthly Reanalysis Maps. Available online: http://ccireanalyzer.org/reanalysis/monthly_maps/index.php (accessed on 24 September 2018).

25. EnergyPlus. Weather Data by Region. Available online: https://energyplus.net/weather-region/europe_wmo_region_6 (accessed on 18 December 2018).

26. ISAVIA. Monthly Report Traffic Statistics/Summary. Available online: https://www.isavia.is/media/1/11-2018-tolur-fyrir-vefsiduna.pdf (accessed on 6 December 2018).

27. Reykjavík Weather Statistics. Available online: www.vedur.is/Medaltalstoflur-txt/Stod_001_Reykjavik.ArsMedal.txt (accessed on 12 October 2018).

28. Stathopoulos, T. Wind and comfort. In Proceedings of the 5th European & African Conference on Wind Engineering—EACWE, Florence, Italy, 19–23 July 2009; pp. 1–16.

29. Neufert, E. *Navrhování Staveb*, 2nd ed.; ConsultinvestInterna: Praha, Česká republika, 2000; ISBN 8090148662.

30. OpenFOAM: The Open Source CFD Toolbox. Available online: https://www.openfoam.com/documentation/user-guide/ (accessed on 3 February 2018).

31. Paul, R.; Dalui, S.K. Wind effects on 'Z' plan-shaped tall building: A case study. *Int. J. Adv. Struct. Eng.* **2016**, *8*, 319–335. [CrossRef]

buildings

MDPI

Article

Parametric Creative Design of Building Free-Forms Roofed with Transformed Shells Introducing Architect's and Civil Engineer's Responsible Artistic Concepts

Jacek Abramczyk * and Aleksandra Prokopska

Rzeszow University of Technology, Department of Architectural Design and Engineering Graphics,
Al. Powstańców Warszawy 12, 35-959, Rzeszów, Poland; aprok@prz.edu.pl
* Correspondence: jacabram@prz.edu.pl; Tel.: +48-795-486-426

Received: 22 February 2019; Accepted: 3 March 2019; Published: 6 March 2019

Abstract: The article concerns a parametric description of unconventional building forms roofed with folded sheeting transformed elastically into shells. The description supports the designer in the search for attractive forms and a rational use of materials. The adoption of strictly defined sets of initial parameters determines the diversification of the designed architectural free-forms. An impact of selected proportions between these parameters on these forms is illustrated by an example of a single structure. Folded elevations and a segmented shell roof make each such structure internally coherent and externally sensitive. The mutual position and proportions of the shape of all elements, such as the roof, eaves, and façades, along with regular patterns in the same structure, determine this consistency of its form and sensitivity to harmonious incorporation into the natural or built environments. The study is a new insight into shaping free-forms of buildings in which the modern and ecological materials determine the important shape and mechanical limitations of these forms. With a skillful approach, the materials allow their extensive use in buildings. However, various interdisciplinary problems related to architectural shaping of free-forms and static and strength work thin-walled shell sheeting roofs must be solved. For effective design it is necessary to use relevant software applications, where spatial reasoning is crucial for ordering the three-dimensional space by means of simplified engineering models.

Keywords: building free-form structure; corrugated shell roof; architectural form; thin-walled open profile; shape transformation; folded sheet; steel construction

---

1. Introduction

Doubly curved shell roofs have been used since the Gothic and became very popular in the Renaissance owing to their attractive architectural forms, strong and stable constructions and big internal, column-free space in buildings [1,2]. To strengthen the shell roofs and improve their stability, complete shells are combined into a single internal coherent shell structure in which ribs are used along the common edges of their adjacent individual shells [3]. These ribs cause the resulting shell structure to be more resistant to building settlement and shocks in the ground [4], with a simultaneous increase of the span of their roof compared with single shells. The use of flat or shell elements made of laminated glass and reinforced polymer as structural members together with metal ones diversifies and improves the attractiveness of the architectural forms of buildings [5,6]. In this way, buildings become friendly to stay, easy to treat [7], more ecological and sensitive to build and natural environments [8,9].

The search for new ways to increase the span of roof shells resistant to diverse external influences, including snow and settlement, determines the development of more and more original and complex types of structural systems and roof coverings [10–14]. The authors' interests are complex roof structures created as a result of connecting many single shells into a single strip made of nominally flat, thin-walled folded sheets transformed elastically into shell forms by supporting them on roof directrices (Figure 1) [15,16].

(a)   (b)

Figure 1. Transformed sheetings: (**a**) an erected complex roof structure of many hyperbolic paraboloid shells and (**b**) an experimental folded shell.

The designer ought to exploit the fact that folds have open profiles and they can easily adapt their deformed shell shape to the shape and mutual position of the roof directrices in a fairly wide range of transformation despite the complicated deformation of their walls causing considerable deformations and rotations of their cross-sections (Figure 2a). The folded structure of such a shell allows all folds after transformation to be maintained straight [17], so they are modeled in a simplified manner to engineering developments by means of ruled surfaces characterized by straight rulings t_i and the line of striction s (Figure 2b) [18]. The rulings t_i model determines the longitudinal axes of all folds in a shell. The line of striction s represents the contraction of all shell folds.

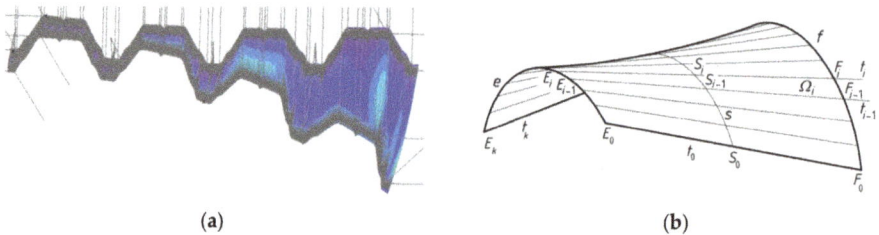

(a)   (b)

Figure 2. Two models used in free-form transformed roof modeling: (**a**) an exact, thin-walled computational model of a nominally plane folded sheet transformed into a shell shape and (**b**) a general warped surface Ω_i stretched over directrices e, f, and described by finite number of rulings t_i.

Prior to transformation the longitudinal axes of all folds in flat sheeting are straight lines parallel to each other and contained in one plane. During the transformation, they occupy strictly defined positions of skew straight lines, where the distances between them slightly and gradually decrease. The width of each transformed fold changes along its length, so that the fold contracts at half-length and expands at the opposite transverse ends. The longitudinal edges of each inner fold in the transformed shell are also skew straight lines. However, the longitudinal edges of the edge folds of the shell bend outwards (Figure 2a) and must be stiffened with hot-rolled profiles or hollow sections [19].

The roof directrices support the shell folds at their ends transversally to the directions of the folds. The shape and mutual position of roof directrices can be adopted relatively arbitrarily, so quite diverse forms of single, thin-walled roof shell sheeting can be achieved (Figure 3) [20].

Figure 3. Two transformed free-form roof structures composed of a few warped shells: (**a**) an amphitheater on the boulevards and (**b**) a number of stores.

In order to obtain effective forms of all folds in the transformed shell, the fold's cross-sections are provided with free lateral deformation when assembling these folds into the shell directrices. Allowing such freedom for all shell folds enables the designer to optimize their initial effort to the lowest possible level. In this way, the folds change their widths and heights so that they contract at their half-length and equalize their positive width increments at their opposite transverse edges, i.e., along both transverse edge directrices. In the computation and modeling the shape of each fold, two conditions are used to optimize the form. One of them determines the location of the aforementioned contraction halfway along the length of each shell fold. The other one preserves the equality of the surface areas of two smooth models created for the same fold before and after its transformation [21].

As a consequence of both conditions, the initial stresses resulting from the fold's shape transformations can be designed as smallest possible, which allows the use of the shell folds as structural elements transferring dead and live loads onto the roof directrices. For this reason, thin-walled folded sheets are readily used for roofing. The skillful character of the shape transformations and freedom in defining roof directrices' shapes can improve the attractiveness of the designed architectural forms [22] and their sensitivity to a harmonious incorporation into the expected natural or built environments [23].

During the shape transformations, shell folds are subjected to transverse bending and twist about their longitudinal axes along the length. In the case of curvatures used in shell roofs, the transverse bending of the folds does not significantly affect the level of their effort. The key is the twist, which, for engineering purposes, is assumed to be constant for the folds whose length ranges from 4 m to 15 m and height from 60 mm to 160 mm [16,24]. In fact, the unit twist angle varies along the length of each twisted fold so that the greater the length of the shell folds the smaller the maximum recoverable unit twist angle that depends on the local elastic instability of the thin-walled flanges and webs of the fold. Therefore, for the cases of large and medium spans of roofs, individual shells are set together into roof structures (Figure 4) and appropriate structural systems dedicated to them must be designed to stiffen these structures and entire buildings [25–31]. Additionally, in order to make the building's form more attractive, folded elevations, inclined to the vertical, are designed (Figure 4) [22].

The simplest way of creating such complex general free-forms of buildings is obtained by combining several complete free-forms, so that their selected walls are contained in common planes (Figure 4) [25,32]. As a result, a single building structure roofed with a structure of few complete shells separated by edges (Figure 1a and Figure 3a) or additional areas completed with other covering elements (Figure 3b), for example glass plates (Figure 4), is obtained.

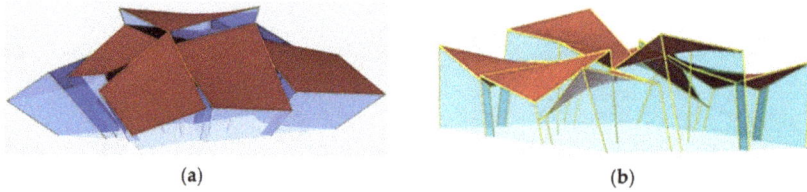

Figure 4. Two different building free-form structures characterized by (**a**) straight roof directrices and (**b**) curved roof directrices.

The interdisciplinary problems discussed and the complexity of their solutions require the designer to have professional knowledge and skills in the field of building, architecture, constructions and spatial reasoning in shaping accurate, and simplified engineering models and placing them in the Euclidean space [33]. In order to support the designer's activities in shaping the discussed free-forms and their structures, the authors' main challenging tasks are reduced to the development of a parametric description of the shaping process of the aforementioned building forms [25] and structural systems intended for them (Figure 5). The developed description allows the authors to create their own computer applications, and in the future tutorials and guidebooks supporting the process of shaping the considered free-forms.

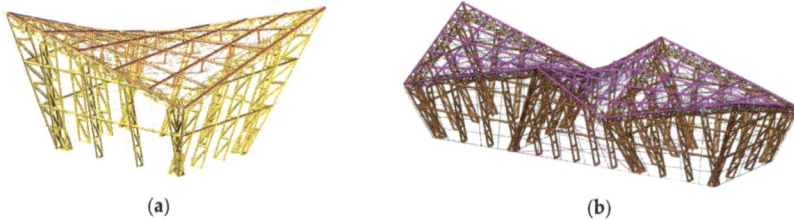

Figure 5. Structural systems of (**a**) a complete free-form building and (**b**) a structure of three such complete buildings.

In present article, the activities of shaping such structures are limited to the parametric description of single and complex forms of general buildings and their roof structures. The description is oriented towards seeking for proportions defining various attractive, consistent general forms of roofs and façades. However, it is not focused on searching for different types of such forms and structures depending on the adopted sets of parameters. This different way is discussed by Abramczyk [21]. The description also takes into account the results of experimental tests [15,16] and computer analyses [34,35] of shape changes and static strength work of all folds in such transformed shells (Figure 6).

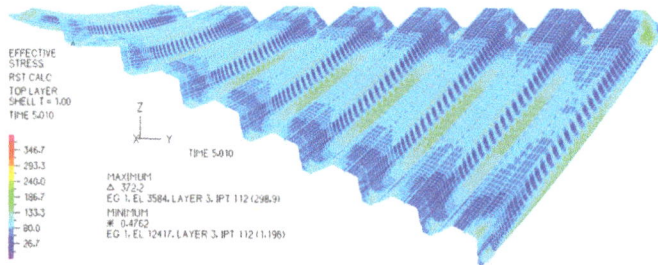

Figure 6. A static strength work of an accurate thin-walled folded computational model of shell sheeting under a characteristic load.

2. Critical Analysis

Forced shape transformations of folded sheets were accomplished by Gergely, Banavalkar, and Parker [36] in the 1970s to create shallow hyperbolic paraboloid roofs and their structures named hypars. These sheet's transformations are ineffective because additional forces acting transversally towards the fold's longitudinal axes have to be used to reduce the widths of all shell folds to the arbitrary or poorly calculated lengths of the fold's supporting lines along roof directrices [37,38]. Various configurations of hypar unit structures are proposed by Bryan and Davies [39].

In the 1990s, Adam Reichhart started shaping corrugated steel sheeting for shell roofing, where all folds underwent big transformations into shell shapes. The transformations were named by Reichhart free deformations [19]. He developed a simple method for geometrical and strength shaping of such shell roofs [15]. According to his method, each roof shell sheet is modeled with a central sector of a right hyperbolic paraboloid limited by a spatial quadrangle [40].

Jacek Abramczyk used the general Reichhart's concept and created a new, more accurate method for shaping the transformed shell roofs. In addition, he assumed that the great freedom in shaping diversified transformed shell forms for roofing can be used to integrate the entire building free-form (Figure 3). Consequently, he decided to incline and fold elevation walls to the vertical depending on the shape of shell roof and entire building [18] (Figures 1 and 2). He noticed that there is interdependence between the efficiency of the roof sheeting transformation and the location of its contraction along the length of each roof shell fold [17]. This relation strongly affects the attractiveness of the entire form and the integrity of the shapes of the roof and elevation [22].

Aleksandra Prokopska conducted multivariate interdisciplinary analyses of some consistent morphological systems that can be designed in harmony with the natural or man-made environments [40,41]. Her research involves many interdisciplinary topics needed to develop experience in shaping various attractive architectural free-forms [24,42].

Some main principles of shaping complete and compound innovative free-forms are the result of the cooperation between Aleksandra Prokopska and Jacek Abramczyk [22]. On the basis of these principles they invented a preliminary version of a method for parametric shaping of the aforementioned folded plane and shell forms [23].

3. Aims and Scope of the Paper

The main aim of the article is to present an innovative parametric description for shaping unconventional free-form building structures composed of a few complete free-forms positioned in one row and connected to each other with selected common plane elevation walls. This description should take into account the fact that in each individual form of such a structure the roof and façade forms should be integrated, which can be obtained by establishing appropriate relations between the dimensions, mutual position, and orientation of their main elements, such as edges and planes. These actions should be guided by the effort to obtain the smallest possible set of parameters that allow an intuitive prediction of approximate proportions between the dimensions of the characteristic elements of the roof and elevation of the subsequent individual forms in the structure under consideration.

In addition, the description should allow the modification of the shape and mutual position of all characteristic edges of the designed structure in order to obtain the possibility of shaping complex folded roof and façades. This action lets the designer conduct a creative search for very diverse unconventional forms characteristic of their current engineering or architectural activity and sensitive to the surrounding environments.

At the same time, the proposed description should allow for the

- adoption of a reference parameter by means of which the expected proportions between elements of the structure, determining its internal consistency, attractiveness, and external sensitivity to the surrounding environment can be defined;

- achievement of effective, i.e., optimal, shapes of all folds in the transformed roof shell in order to reach the smallest possible values of the effort of these folds and large deformations of the fold's walls allowing for relatively high curvature of the roof shell;
- fairly free adoption of the shape and position of roof directrices, which decisively affects the attractiveness of the shell roof form, the entire single free-form, and even the entire structure as well as its external sensitivity;
- development of some original parametric computer applications supporting the designer's activities in the field of creating simplified models of shell roofs and entire structures; and
- development of parametric models for structural systems intended to the discussed types of the structures, in the future.

4. The Concept and the Range of the Work

In the parametric description presented in the sections that follow; the solutions of the issues appearing in the shaping process of different free-forms of buildings covered with folded roofs transformed to the shell form are presented. The actions and objects required for the above solutions are discussed on an example of shaping one single free-form and one structure of three free-forms.

In the first part of the description, a method for integrating the free-forms of roofs and façades by means of flat figures called reference polygons is proposed. The integration is achieved using a possibly small set of parameters defining some approximately selected values and proportions between the basic dimensions of these forms. Two separate plane parts of this polygon, corresponding to the spatial forms of roof and façade, are defined. At the same time, the polygon defines the slopes of the characteristic edges and planes of the roofs and façades to the horizontal base plane.

In the second part of the description, the aforementioned reference polygon P_r is expanded into a solid Σ, inter alia based on straight lines perpendicular to the plane of the polygon P_r. A very important role in shaping the final form of the sought-for structure is played by the closed spatial quadrilateral eaves consisting of four straight sections and four planes of façade walls inclined to the vertical. At this step, it is possible to modify two opposite eaves lines to the form of curved directrices of the roof shell.

In the third part of the parametric description further complete free-forms are defined, of which the resulting structure will be composed. In this article, single forms must have flat walls and a flat horizontal base contained in one initially adopted basic plane. The roof shell form is irrelevant to the above process of combining individual forms.

In the fourth part of the description, the created single free-forms are joined by common elevation walls in a single-row structure (Figure 7). In the presented example, the roof consists of three single shells joined into one continuous structure, where the complete shells are separated by common edges disturbing the smoothness of the entire roof structure.

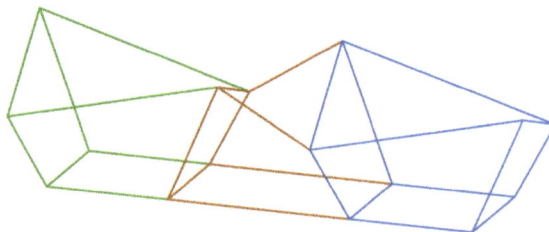

Figure 7. A structure composed of three complete free-forms.

In the fifth part, the presented parametric description allows modifying the forms of the roof and façade of the achieved structure in a certain, but not very large, range. The significant effect of this modification is a folded plane-walled façade structure and multisegment roof structure. At this step, the thickness and overhang of the roof shell are defined.

In one of the parts of the above description, a regular elevation pattern must also be defined as the sum of the appropriately arranged flat strips of the façade glass panels. This pattern must be integrated with the general form of the structure and its individual elements.

Finally, visualization of the example structure is presented, which confirms the correctness of the assumptions, actions, and objects used in the discussed parametric description, and the wide possibilities of obtaining various free-forms and their structures. This visualization also illustrates that the structures built on the basis of this description have the internal integrity of the form of their elements and the external sensitivity to the natural environments. The method of creating the visualized structure is presented in the subsequent sections of the article.

5. Intuitive Parametric Shaping of General Forms of Free-Form Buildings

In order to design a general building free-form using the proposed method, the following action should be performed. The values of the following parameters describing transverse dimensions of the designed free-form are adopted. Eight parameters: $a_1 = |C_1C_2|$, $b_1 = |C_2C_3|$, $c_1 = |C_5C_{6r}|$, $d_1 = |C_4C_5|$, $a_2 = |C_1C_{15}|$, $b_2 = |C_{15}C_{17}|$, $c_2 = |C_6C_{11}|$, and $d_2 = |C_{10}C_{11}|$ determine four auxiliary rectangles Δ_i ($i = 1$ to 4) (Figure 8a). For example, $\Delta_1 = C_1C_2C_{14}C_{6r}$ and $\Delta_2 = C_4C_5C_{6r}C_{7r}$. The rectangles Δ_i are located in a very specific manner so that each of Δ_i has one edge contained in the z-axis. The rectangles Δ_1 and Δ_4 are the initial approximation of the elevation form. The Δ_2 and Δ_3 rectangles are the initial approximation of the roof form.

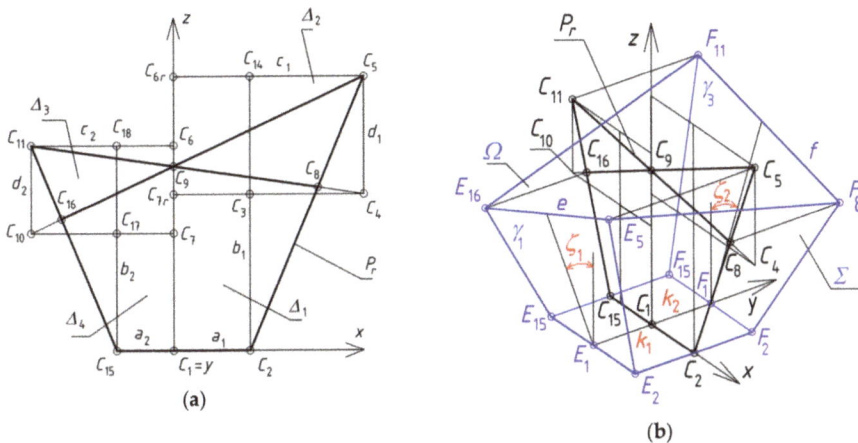

Figure 8. (a) An auxiliary flat reference polygon P_r. (b) A spatial free-form Σ.

Two rectangles—Δ_1 and Δ_4—have the common vertex C_1 located at the origin of the adopted orthogonal coordinate system [x, z]. Points C_8 and C_{16} are the intersection of the respective segments $C_4C_{11} \cap C_2C_5$ and $C_5C_{10} \cap C_{11}C_{15}$.

On the basis of the sum Δ of the above four rectangles Δ_i, a plane reference polygon P_r is built, so that its vertices are identical with some selected rectangle's vertices. These vertices are as follows: C_1, C_2, C_5, C_8, C_9, C_{11}, C_{15}, and C_{16} (Figure 8a). The shape integration of the roof and elevations is ensured by means of the proper proportions between the areas of triangles $C_8C_5C_9$ and $C_9C_{16}C_{11}$ and quadrangles $C_1C_2C_8C_9$ and $C_1C_9C_{16}C_{15}$ of P_r. These proportions express the real proportions between the size of the roof shell and elevation wall under consideration, and between the inclination of the eaves and elevation walls to the vertical. Thus, in this step of the presented description, these eight parameters should be employed.

In the next step, the reference polygon P_r is extended into a spatial free-form Σ sought (Figure 8b). Therefore, six straight lines perpendicular to the plane of P_r should be passed through the vertices: C_2, C_5, C_8, C_{11}, C_{15}, and C_{16}. Vertices E_2, E_5, E_{16}, F_2, F_8, F_{11}, and F_{15} of Σ are determined as the points of the intersection of the above six straight lines and two planes γ_1 and γ_3 of two gable walls of Σ. In this step four additional parameters are employed: two parameters—ζ_1 and ζ_2—defining the dihedral angles of the inclination of planes γ_1 and γ_3 to the vertical, respectively, and two parameters—k_1 and k_2—defining the distance of E_1 and F_1 from the plane (x, z) (Figure 8b).

To define the above figure Σ by means of one independent variable a, the main proportions, included in Table 1, are adopted. In addition, the last two parameters are $\mathrm{tg}(\zeta_1) = 0.429$ and $\zeta_2 = \zeta_1$, where ζ_1 – measure of the angle between γ_1 and (x, z) and ζ_2 – measure of the angle between γ_3 and (x, z).

Table 1. Parameters adopted for the examined architectural free-form shown in Figure 8.

Proportion	Value
a_1/a	1.00
b_1/a	2.00
c_1/a	2.50
d_1/a	1.50
k_1/a	1.00
a_2/a	0.75
b_2/a	1.50
c_2/a	1.875
d_2/a	1.125
k_2/a	1.0

[1] a = 5000 in millimeters.

The coordinates of its characteristic points are given in Table 2, and the visualization of the architectural form related to the discussed general form is shown in Figure 9. The aforementioned proportions ensure that this configuration of Σ meets the expectations of the visual attractiveness of the designed free-form and its internal integration.

Table 2. Coordinates of the vertices of the examined general free-form shown in Figure 8.

Vertex	X-Coordinate	Y-Coordinate	Z-Coordinate
E_2	5000.0	−5,000.0	0.0
E_{15}	−5000.0	−5,000.0	0.0
F_2	5000.0	5000.0	0.0
F_{15}	−5000.0	5000.0	0.0
E_5	12,500.0	−12,500.0	17,500.0
E_{16}	−7747.9	−8533.1	8243.8
F_8	9471.1	9471.1	10,432.7
F_{11}	−9375.0	10,625.0	13,125.0

[1] values in millimeters.

The obtained spatial quadrangle $E_5F_8F_{11}E_{16}$ represents the eaves of the roof shell. Two opposite segments E_5E_{16} and F_8F_{11} of the aforementioned quadrangle (Figure 8b) are called roof directrices e and f. The directrices are here two straight lines.

In the end, an elevation pattern is defined by means of two families of straight lines (Figure 9). The parameters describing the pitch, position and orientation of the elevation pattern should be defined. In the presented examples, only two parameters defining the pitch in the horizontal and vertical direction are employed. The roof thickness and cantilever protruding out of the elevation planes are two other parameters that affect the attractiveness, internal integration of the building form and external integration of the form with the built or natural environment.

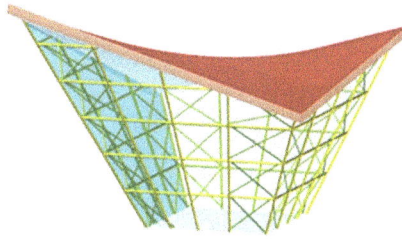

Figure 9. Architectural study of the free building form roofed with transformed shell roof.

6. Elevation Shaping

Regular elevation patterns together with general shapes and proportions between the main parts of the free-form buildings like elevations, roofs, and structural systems play an important role in improving the visual attractiveness of these forms and their harmonic integration with man-made environments. These patterns are designed by means of visible elements supporting elevation glass plates or filling the areas between these plates. The pitch, orientation, and position of these elements should be analyzed. Members of structural systems may also be the visible elements forming the regular pattern.

The models presented below take account of the thickness of the roof and regular pattern of elevations (Figure 10). The main object of this step is a model taking account of the aforementioned properties of roof and elevations. They are sufficiently accurate in engineering developments.

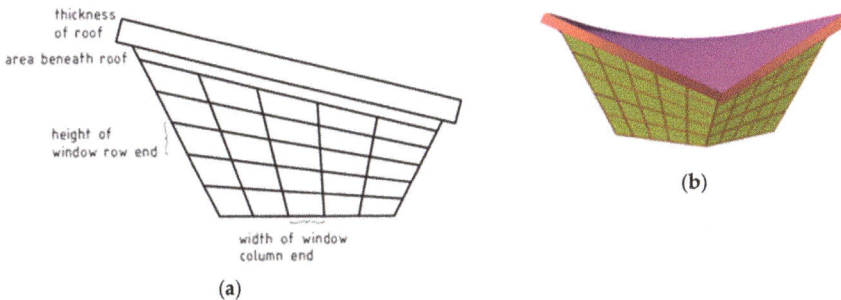

Figure 10. Constant height of the window row ends along the oblique elevation wall edges and constant width of the subsequent windows columns along the eaves and base in the horizontal direction: (**a**) a scheme of an elevation wall and (**b**) a model of a free-form.

Its surfaces, planes, lines, and points are the main auxiliary geometrical elements. Intersections, displacements, and rotations of these elements allow all building elements as finite sectors, sides, and edges to be modeled and arranged relative to the building construction axes.

The mutual location of the roof and elevations results from the structure and overall dimensions of the general building form. Therefore, the new parameters used in this step are

- u—the thickness of roof,
- v—the roof overhang outside the outline of elevation walls, and
- w_1, w_2—the pitch (may also be position and inclination) of regular elevation pattern in horizontal and vertical direction.

The range of possible regular elevation patterns considered in the present paper is limited to two configurations shown in Figures 10 and 11. This is due to the main aim of the paper concerning the use of the parametric description for creative shaping innovative consistent spatial building free-forms and their structures relevant to the built and natural environments.

The above first pattern considered (Figure 10) is reduced to only one type formed from five strips of glass plates positioned in the direction close to the vertical and five strips located in the direction close to the horizontal base plane. The presented pattern forms a network of quadrilateral plane glass "cells" changing equally their widths and heights.

The thickness of the roof shown in Figure 8, together with the unfilled with glass elevation part located beneath the roof create a strip whose height is also important in relation to the dimensions of the roof and elevations, which also affects the process of architectural shaping. The constant division of elevation areas with the orthogonal or, in particular, the diagonal lines of the pattern may produce a fine impression of harmony and integration of all elements and the entire free-form. The compatible constant or variable changes of very often improve the aforementioned features of the designed building.

- the height of each window row together with the change of the elevation height and
- the width of each window column together with the change of the elevation width

The second configuration of the elevation pattern is presented in Figure 11, where the widths of the subsequent horizontal and vertical strips change. The size of the horizontal strips changes in the vertical direction. The size of the vertical strips changes in the horizontal direction. Both changes can positively affect the user's wellbeing.

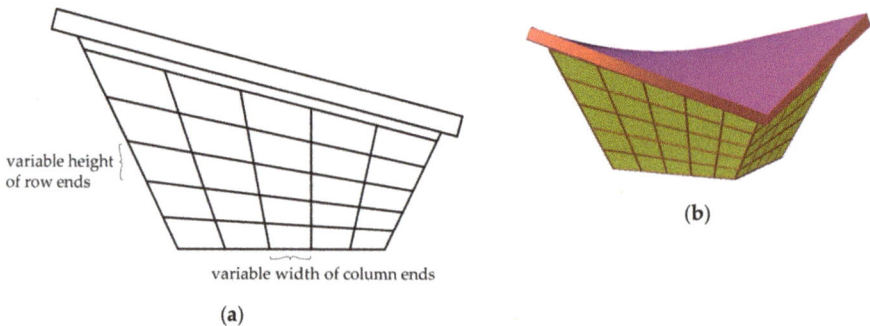

Figure 11. Linear variation of the height of the window row ends along the oblique elevation wall edges and constant width of the subsequent windows columns along the eaves and base in the horizontal direction: (**a**) a scheme of an elevation wall and (**b**) a model of a free-form.

Thin elements between glass plates can perform an important function improving the visual attractiveness of the elevations from the outside. The structural system can be gently truncated inside of a building to enhance the internal attractiveness. If it is visible by a user from the inside of the building, then the user's comfort of wellbeing can be improved. It can also be translated outside the building (Figure 12), which can change radically the user's visual impression. Structural systems and regular elevation patterns may be designed independently so that each can perform different functions: structural or visual by the observation from the inside or outside of the building designed.

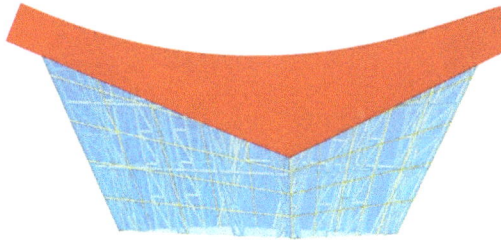

Figure 12. Visualization of a free-form building with visible structural system.

7. Attractive Shape Proportions for Free-Forms

To obtain satisfactory forms of buildings it is necessary to analyze some proportions between the size, orientation and inclination of all main elements of the designed building such as roof, elevations, and their elements like eaves and edges. Attractive complete forms characterized by integral forms of the roof, elevations, and eaves can be developed as the result of assuming the strictly defined relationships between the discussed parameters, for example the ones included in Tables 3 and 4, which leads to the reference polygon shown in Figure 13. The properties are adopted according to the guidelines and functional dependence defined by Abramczyk [22]. The result of the above dependencies is that the sum of four rectangles Δ_i is a flat figure Δ symmetrical towards the vertical z-axis. The reference polygon P_r, created on the basis of Δ, is also z-axis-symmetrical. The architectural free-form shown in Figure 12 was achieved employing the aforementioned procedure, parameters from Tables 3 and 4 and tg ($\zeta = \zeta_1 = \zeta_2$) = 0.25.

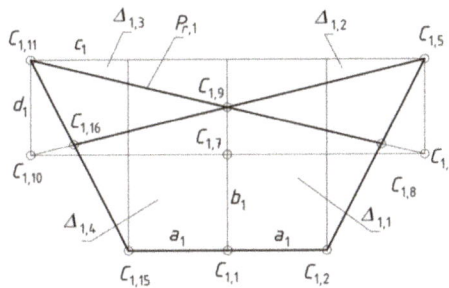

Figure 13. Symmetrical reference polygon P_r.

The parameters $v = |M_1B_1| = |M_2B_2| = |M_3B_3| = |M_4B_4| = |N_1B_1| = |N_2B_2| = |N_3B_3| = |N_4B_4| = |D_{d1}D_{g1}| = |D_{d2}D_{g2}| = |D_{d3}D_{g3}| = |D_{d4}D_{g4}|$ and $u = |D_{d1}D_{g1}|$, included in Table 3, determine the overhang and thickness of the shell roof (Figure 14). To achieve attractive proportions, only nine parameters—$a_1, b_1, c_1, d_1, k_1, \zeta_1, u, v, w_1 = w_2$—are needed in the case of shaping symmetrical free-forms. The width of the elevation walls is irrelevant in the present example.

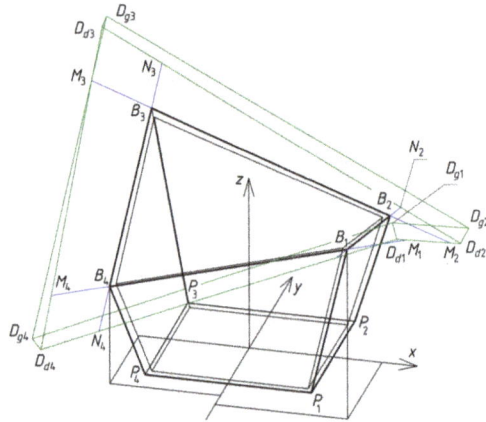

Figure 14. Simplified model of a general free-form taking account of the roof thickness and overhang end the elevation wall thickness.

In addition, the roof shell can be translated in the vertical direction as was done for the case of the free-form shown in Figure 15 and the structure presented in the next section. The coordinates of the characteristic points of such architectural form shown in Figures 13 and 15 are given in Table 5.

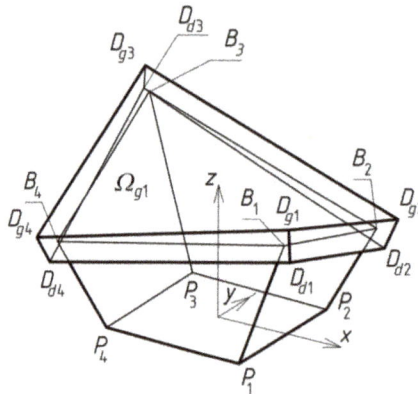

Figure 15. Simplified model of the discussed general free-form taking account of the roof thickness and overhang.

Table 3. Parameters adopted for the examined architectural free-form from Figures 13 and 15.

Proportion	Value
a_1/a	1.00
b_1/a	1.00
c_1/a	2.00
d_1/a	1.00
k_1/a	1.00
v/a	0.16
$w_1 = w_2$	5
u/a	0.4

[1] a = 5000 in millimeters.

Table 4. Coordinates of the vertices of the examined architectural free-form shown in Figures 13 and 15.

Vertex	X-Coordinate	Y-Coordinate	Z-Coordinate
$C_{1,1}$	0.0	0.0	0.0
$C_{1,2}$	5000.0	0.0	0.0
$C_{1,5}$	10,000.0	0.0	10,000.0
$C_{1,8}$	7777.7	0.0	5555.6
$C_{1,9}$	0.0	0.0	7500.0
$C_{1,11}$	−10,000.0	0.0	10,000.0
$C_{1,15}$	−5000.0	0.0	0.0
$C_{1,16}$	−7777.7	0.0	5555.6

[1] values in millimeters.

Table 5. Parameters achieved for the examined architectural free-form from Figure 15.

Vertex	X-Coordinate	Y-Coordinate	Z-Coordinate
P_1	5000.0	−5000.0	0.0
P_2	5000.0	5000.0	0.0
P_3	−5000.0	5000.0	0.0
P_4	−5000.0	−5000.0	0.0
B_1	10,000.0	−7500.0	10,000.0
B_2	7777.7	6388.9	5555.6
B_3	−10,000.0	7500.0	10,000.0
B_4	−7777.7	−6388.9	5555.6
D_{g1}	10,650.6	−8016.9	11,382.6
D_{g2}	8697.0	7448.5	5996.4
D_{g3}	−10,650.6	8016.9	11,382.6
D_{g4}	−8697.0	−7448.5	5996.4
D_{d1}	11,129.5	−8585.8	9526.0
D_{d2}	8156.6	6739.5	4206.1
D_{d3}	−11,129.5	8585.8	9526.0
D_{d4}	−8156.6	−6739.5	4206.1

[1] values in millimeters.

The calculation of the supporting conditions for the subsequent folds in the example shell, shown in Figure 15, should be carried out according to the algorithm proposed by Abramczyk [21] in [21]. On the basis of the supporting conditions obtained, the lengths of the supporting lines for the subsequent folds along the roof directrices should be calculated using the computer application written by Abramczyk [22] discussed in [21].

Some functional dependencies between the above parameters can be a basis for the estimation of the innovation and attractiveness of these forms to be designated. The proportions should result in the following;

- the inclination of each elevation wall to the vertical relates to the slope of all eaves to the horizontal base,
- the sizes of roof and elevation are close to each other,
- the expected proportions between the height and width of the building as well as the roof's thickness along eaves ought to be maintained,
- the curvature and contraction of the roof shell correspond to the building's height measured along each of four oblique elevation edges, and
- the shape of the visible part of the building's structural system relates to the elevation pattern.

8. Compound Free-Form Structures

The proposed parametric description also concerns creating complex structures composed of many individual forms with shared elevation walls. Few individual free-forms can be arranged in one row or in rows and columns (Figure 4) [23]. The structures investigated in the present article are arranged only in one row. They are determined by means of inseparable symmetrical reference polygons $P_{r,i}$ created subsequently in accordance with the algorithm described previously for single P_r.

The sum Σ of such three complete forms Σ_i (i = 1 to 3) having common planes of the elevation walls and arranged in one direction is presented in Figure 16. The actions related to the construction of form Σ start with the determination of three auxiliary flat reference polygons $P_{r,i}$ (i = 1 to 3) placed in the same vertical plane.

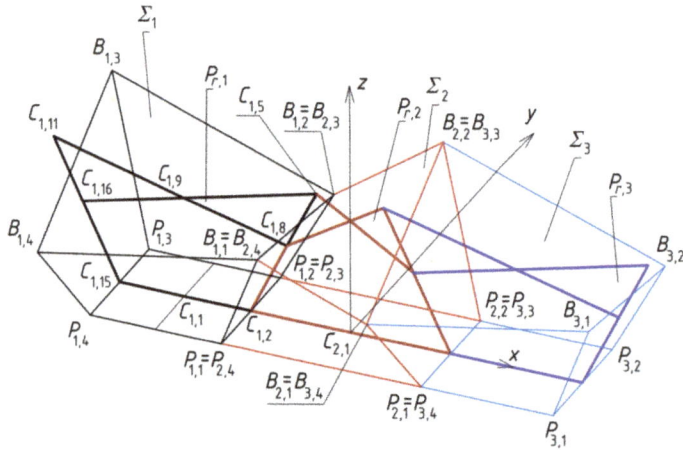

Figure 16. An edge model of three inseparable complete forms Γ_i created on the basis of three reference polygons $P_{r,i}$.

According to the proposed algorithm, $P_{r,1}$ is created on the basis of four parameters: a_1, b_1, c_1, and d_1 defining four rectangles $\Delta_{1,j}$ (j = 1 to 4) (Figure 17). Next, the value of the parameter a_2 should be adopted in order to construct point $C_{2,1}$ on the extension of the straight line ($C_{1,2}$, $C_{1,15}$). Point $C_{2,17}$ is constructed at the intersection of the line passing through $C_{2,1}$ and perpendicular to the straight line ($C_{1,2}$, $C_{2,1}$) with the straight line ($C_{1,11}$, $C_{1,5}$). Because $C_{2,11}$ is similar to $C_{1,5}$, the value of parameter c_2 is calculated as the length of section $C_{2,11}C_{2,17}$. In addition, $C_{2,10}$ is the intersection of the straight line passing through $C_{2,5}$ and $C_{2,16}$ = $C_{1,8}$ with the straight line ($C_{1,5}$, $C_{1,4}$). Point $C_{2,5}$ is symmetrical to $C_{1,5}$ about the axis ($C_{2,1}$, $C_{2,17}$). The value of parameter d_2 can be calculated as the length of segment $C_{2,10}$, $C_{2,11}$. Rectangles $\Delta_{2,3}$ and $\Delta_{2,4}$ can be created on the basis of the aforementioned points. The last parameter is $b_2 = b_1 + d_1 - d_2$.

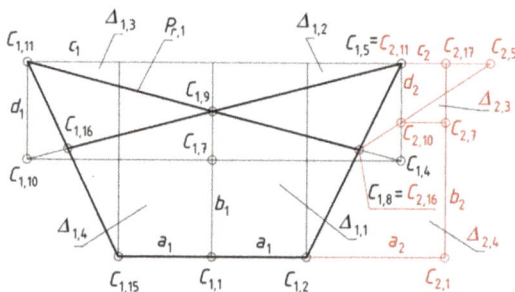

Figure 17. Creation of reference polygon $P_{r,2}$.

Parameters a_2, b_2, c_2, and d_2 define both rectangles $\Delta_{2,3}$ and $\Delta_{2,4}$ and rectangles $\Delta_{2,1}$ and $\Delta_{2,2}$, which are symmetrical to the previous ones (Figure 18). Summing up, the value of parameter a_2 should be adopted, whereas the values of c_2, b_2, and d_2 have to be calculated.

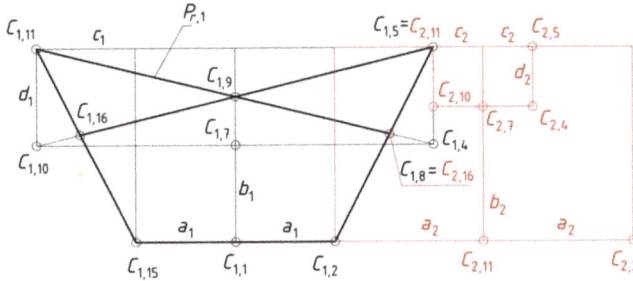

Figure 18. Creation of reference polygon $P_{r,2}$.

On the basis of rectangles $\Delta_{2,i}$ ($i = 1$ to 4), including their vertices, polygon $P_{r,2}$ is constructed (Figure 19). Polygons $P_{r,1}$ and $P_{r,2}$ have one common side $C_{1,2}C_{1,5}$.

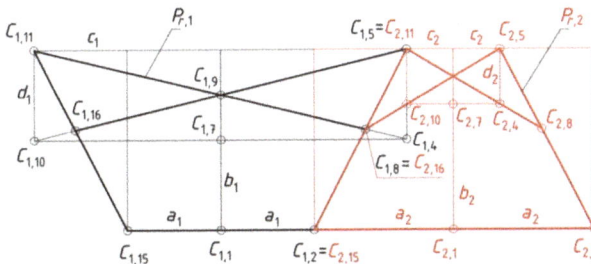

Figure 19. Geometrical properties of the reference polygon $P_{r,2}$.

In the same way as $P_{r,2}$, the last polygon, $P_{r,3}$, is created (Figure 20). The creation of $P_{r,3}$ requires adopting the value of a_3 and calculating the values of b_3, c_3, and d_3.

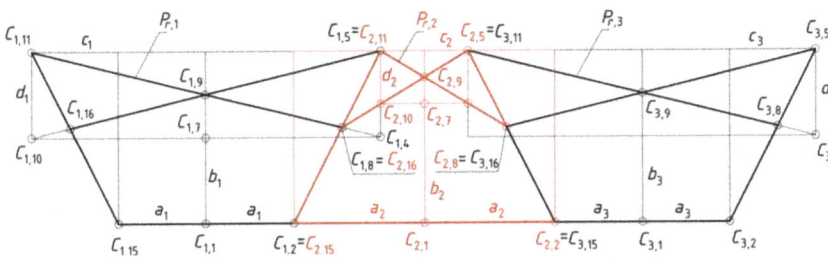

Figure 20. Creation of the reference polygon $P_{r,3}$.

The coordinates of the selected characteristic points $C_{i,j}$ (for $i = 1$ to 3 and $j = 1$ to 16) of the discussed complex reference polygon $P_r = \Sigma\, P_{r,i}$ shown in Figure 20 are given in Table 6. The parameters used are $a_1 = a_3 = a$, $b_1 = b_3 = a$, $c_1 = c_3 = 2a$, $d_1 = d_3 = a$, where the values of the main parameters a, b, c, d are given in Table 1. In addition, $a_2/a = 1.5$ is adopted and $b_2/a = 1.385$, $c_2/a = 0.500$, and $d_2/a = 0.615$ are calculated. Based on the whole polygon $P_r = \Sigma\, P_{r,i}$ ($i = 1$ to 3) (Figure 20), a spatial form Σ, which is the sum of Σ_{i_r} is created (Figure 16) using the constructions presented in Section 5. The other parameters are: $k_1 = k_2 = k_3 = a$ and $\zeta_1 = \zeta_2 = \zeta_3 = 14.0362°$.

Table 6. Parameters achieved for the examined architectural free-form from Figures 16–25.

Vertex	X-Coordinate	Y-Coordinate	Z-Coordinate
$C_{1,1}$	−12,500.0	0.0	0.0
$C_{1,2}$	−7500.0	0.0	0.0
$C_{1,5}$	−2500.0	0.0	10,000.0
$C_{1,8}$	−4705.2	0.0	5589.6
$C_{1,9}$	−12,500.0	0.0	7500.0
$C_{1,11}$	−22,500.0	0.0	10,000.0
$C_{1,15}$	−17,500.0	0.0	0.0
$C_{1,16}$	−20,277.8	0.0	5555.6
$C_{2,1}$	0.0	0.0	0.0
$C_{2,2}$	7500.0	0.0	0.0
$C_{2,5}$	2500.0	0.0	10,000.0
$C_{2,9}$	0.0	0.0	8461.5
$C_{3,1}$	12,500.0	0.0	0.0
$C_{3,2}$	17,500.0	0.0	0.0
$C_{3,5}$	22,500.0	0.0	10,000.0
$C_{3,8}$	20,277.8	0.0	5555.6
$C_{3,9}$	12,500.0	0.0	7500.0

[1] values in millimeters.

Finally, the created form Σ is transformed into a complex structure Σ_w which is characterized by folded elevation walls and shell roof structure (Figure 21). This modification of Σ into Σ_w can be done in two ways. The first way is based on displacements and rotations of two opposite gable wall planes of the form Σ_2 and division of the compound shell roof of Σ into three complete shell sectors creating the discontinuous shell roof structure. The displacements and rotations depend on the values of $k_{2,1} < k_{1,1} = k_{3,1}$, and $\zeta_{2,1}$ adopted for the form Σ_2. The value of $\zeta_{2,1}$ may be equal to or different from the values of $\zeta_{1,1}$ and $\zeta_{3,1}$. In the presented example, the discussed structure Σ_w is z-axis-symmetrical: $k_{2,1} = 0.2a$ and $\zeta_{2,1} = 9.0362°$ (Figure 21). The coordinates of the characteristic vertices of this structure, shown in Figures 21 and 22, are given in Table 7.

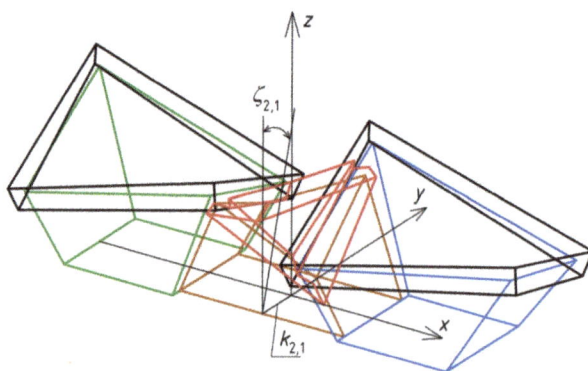

Figure 21. A folded structure of three complete free-forms.

Points $D_{gj,i}$ (for $i = 1$–3, $j = 1$–4) are the vertices of three closed edge lines of the upper surfaces of three complete roof shells Ω_{gi} forming the shell roof structure. Points $D_{dj,i}$ are the vertices of three closed edge lines of the lower surfaces of the same shell roof structure.

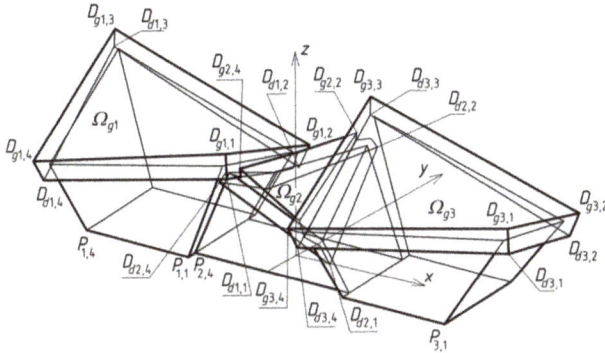

Figure 22. A folded structure of three complete free-forms roofed with multisegment shell structure.

Table 7. Parameters achieved for the examined architectural free-form from Figures 22–25.

Vertex	X-Coordinate	Y-Coordinate	Z-Coordinate
$P_{1,1}$	−7500.0	−5000.0	0.0
$P_{1,2}$	−7500.0	5000.0	0.0
$P_{1,3}$	−17,500.0	5000.0	0.0
$P_{1,4}$	−17,500.0	−5000.0	0.0
$B_{1,1}$	−2500.0	−10,000.0	10,000.0
$B_{1,2}$	−4722.2	−6388.9	5555.6
$B_{1,3}$	−22,500.0	10,000.0	10,000.0
$B_{1,4}$	−20,277.8	−6388.9	5555.6
$P_{3,1}$	17,500.0	−5000.0	0.0
$P_{3,2}$	17,500.0	−5000.0	0.0
$P_{3,3}$	7500.0	5000.0	0.0
$P_{3,4}$	7500.0	5000.0	0.0
$B_{3,1}$	22,500.0	−10,000.0	10,000.0
$B_{3,2}$	20,277.8	6388.9	5555.6
$B_{3,3}$	2500.0	10,000.0	10,000.0
$B_{3,4}$	4722.2	−6388.9	5555.6
$P_{2,1}$	7500.0	−4000.0	0.0
$P_{2,2}$	7500.0	4000.0	0.0
$P_{2,3}$	−7500.0	4000.0	0.0
$P_{2,4}$	−7500.0	−4000.0	0.0
$B_{2,1}$	5327.1	−4648.7	4345.8
$B_{2,2}$	5327.1	5256.5	8418.0
$B_{2,3}$	−5327.1	4648.7	4345.8
$B_{2,4}$	−3291.0	−5256.5	8418.0
$D_{g1,1}$	−1849.4	−8016.9	11,382.6
$D_{g1,2}$	−3803.0	7448.5	5996.4
$D_{g1,3}$	−23,150.6	8016.9	11,382.6
$D_{g1,4}$	−21,197.0	−7448.5	5996.4
$D_{d1,1}$	−1370.5	−8585.8	9526.0
$D_{d1,2}$	−4343.4	6739.5	4206.1
$D_{d1,3}$	−23,629.5	8585.8	9526.0
$D_{d1,4}$	−20,656.5	−6739.5	4206.1
$D_{g2,1}$	3754.3	−5610.5	6043.2
$D_{g2,2}$	2074.9	5579.2	9158.0
$D_{g2,3}$	−3754.3	5610.5	6043.2
$D_{g2,4}$	−2074.9	−5579.2	9158.0
$D_{d2,1}$	4101.8	−5494.0	4638.4
$D_{d2,2}$	3540.5	6390.6	7919.0
$D_{d2,3}$	−4101.8	5494.0	4638.4

Table 7. *Cont.*

Vertex	X-Coordinate	Y-Coordinate	Z-Coordinate
$D_{d2,4}$	−3540.5	−6390.6	7919.0
$D_{g3,1}$	23,150.6	−8016.9	11,382.6
$D_{g3,2}$	21,197.0	7448.5	5996.4
$D_{g3,3}$	1849.4	8016.9	11,382.6
$D_{g3,4}$	3803.0	−7448.5	5996.4
$D_{d3,1}$	23,629.5	−8585.8	9526.0
$D_{d3,2}$	20,656.5	6739.5	4206.1
$D_{d3,3}$	1370.5	8585.8	9526.0
$D_{d3,4}$	4343.4	−6739.5	4206.1

[1] values in millimeters.

For the cases of symmetrical structures, Σ, the aforementioned modification, can be performed by means of stiff motions of three edges $P_{1,1}B_{1,1}$, $P_{1,2}B_{1,2}$, and $B_{1,1}B_{1,2}$ belonging to the shared wall plane of Σ_1 and Σ_2. In the presented example (Figure 23), the symmetry of Σ_2 towards z-axis ensures that the created gable wall of Σ_2 is plane.

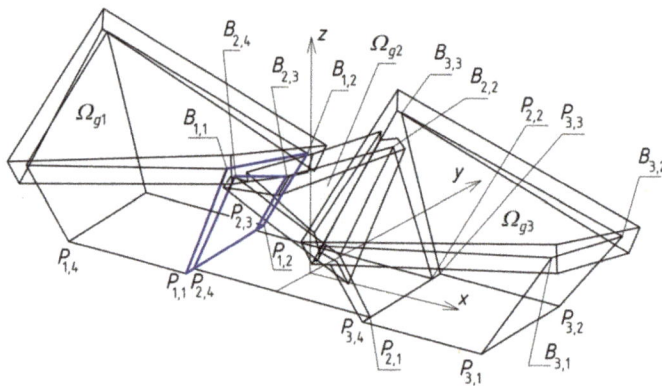

Figure 23. A folded structure of three complete free-forms.

In order to obtain the positions of points $P_{2,4}$, $B_{2,4}$, $P_{2,3}$, and $B_{2,3}$ (Figure 24) the following transformations must be performed. First, translation $T_{1,1}$ and rotation $R_{2,4}$ of the line $(P_{1,1}, B_{1,1})$ is executed. Next, the translation $T_{2,4}$ of point $P_{2,4}$ along the obtained straight line is done to obtain the position of point $B_{2,4}$.

The position of point $B_{2,3}$ can be achieved as a result the intersection of two straight lines $r_{2,3}$ and $r_{2,B}$ determined as follows. To obtain $r_{2,3}$, the straight line $r_{2,P}$ parallel to $(P_{1,2}, B_{1,2})$ and passing through $P_{2,3}$ must be transformed using rotation $R_{2,3}$ around $P_{2,3}$. To obtain $r_{2,B}$, the straight line $r_{2,4}$ parallel to $(B_{1,1}, B_{1,2})$ and passing through $B_{2,4}$ must be transformed using rotation $R_{2,B}$ around $B_{2,4}$. For symmetrical forms, the restriction $R_{2,4} = R_{2,3}$ has to be met in order to obtain flat gable wall of Σ_2. The values of the aforementioned stiff motions are given in Table 8.

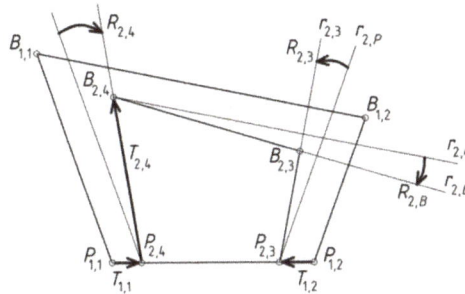

Figure 24. Stiff motions of three edges belonging to one of the walls of Σ_1 resulting in three edges of the corresponding wall of Σ_2.

Table 8. Measures of the stiff motions adopted for free-form shown in Figure 24.

Parameter	Value	Unit
$T_{1,1}$	1000.0	[mm]
$T_{1,2}$	1000.0	[mm]
$T_{2,4}$	9495.2	[mm]
$R_{2,3}$	5.000	[°]
$R_{2,4}$	5.000	[°]
$R_{2,B}$	5.000	[°]

On the basis of the above Σ_w structure, the architectural free-form shown in Figure 25 was visualized. It has an attractive shape and spring and summer colors, which makes it sensitive to the natural environment. The regular pattern of its elevation walls created in the form of thin dark lines evenly distributed on the outer surface and running in two orthogonal directions is marked in a very sensitive way. The coloring of the two spatial lateral free-forms of the complex architectural structure harmonizes with the colors of the natural environment. The yellow color of the minor middle part of the structure refers to the color scheme of the lateral parts of the roof structure and the natural environment. The red color of the middle roof shell stands out from the coloring of the previously mentioned parts and the natural environment.

Figure 25. Visualization of a complex architectural free-form located in the natural summer environmental conditions.

9. Conclusions

The novel parametric description of shaping architectural free-forms of buildings oriented to the shape integration of their roofs and façades is discussed in the article. The description uses the smallest possible set of parameters defining the general forms and regular elevation patterns of these forms. The values of these parameters are the lengths or angles of slopes of the selected roof or façade elements like edges and surfaces.

In the case of symmetrical general forms of buildings, only four parameters {a, b, c, d} describe the geometrical properties of a flat reference polygon, which roughly define the proportions between the size of the roof and façade elements. In order to define the entire general form, the description requires two additional parameters—ζ and k—defining the shape of this form in the direction perpendicular to the direction highlighted previously and also parallel to the building horizontal base. In the case of unsymmetrical general forms of buildings, another six parameters {$a_2, b_2, c_2, d_2, \zeta_2, k_2$} must be adopted.

The possibility of adopting one leading parameter called a reference parameter as only one independent variable defining some basic proportions between selected roof and elevation elements is discussed. The proportions are defined using functions helpful in searching for similar types of interesting architectural free-forms. More comprehensive studies of the aforementioned issues seem to be targeted at an assignment of the ranges in which the values of the selected parameters and their proportions are changed to such groups of free architectural forms characterized by similar properties.

An important role in the aforementioned description is played by the modeling process of the nominally plane folded sheeting transformed into shell shapes, which is based on the shape and mutual position of two roof directrices that can be adopted fairly arbitrarily in the planes of façade walls. On the basis of the directrices, the arrangement and shapes of the folds in the transformed shell, as well as the spacing of the points fixing the transformed fold's ends to the shell directrices, have to be precisely calculated. Because of the relatively complex iterative mathematical calculations combined with the above two conditions, optimizing the fold's shapes it is necessary to use the original application developed by the present authors in the Rhino/Grasshopper program.

The application of the proposed parametric description to create a free-form structure composed of three individual forms connected with shared elevation walls and arranged in a single row has been demonstrated. In this case, the proportions between the dimensions of the roof and façade are defined by three reference polygons contained in one common plane and three groups of parameters. The values of some selected parameters of these groups are taken as identical due to the common base plane and common selected walls of the individual component forms of the created structure. The constructed flat complex reference polygon structure P_r, composed of three reference polygons P_{ri} connected with each other, must be extended to a spatial form in the direction orthogonal to the plane in which it is contained. This action requires the designer to be able to use the spatial reasoning in locating various simplified models of engineering objects in the three-dimensional space. To obtain the integrity of the form of a building the dimensions, the orientation and position of its roof elements, such as eaves, and façade elements, such as the edges of the walls, are interrelated by means of P_r.

Ultimately, the spatial structure can be modified to a form characterized by a folded façade and roof structure in order to make it more sensitive to the existing natural or built environments. Such a modification is presented on an example of a specific structure embedded in the accepted natural environment. The visualization of this structure is shown in one of the presented figures, and its measurement characteristics are included in the respective tables.

It is advisable to develop a method of shaping regular elevation patterns on the façade wall planes, of variable pitch, location, and orientation. The authors intend to develop a concept of shaping shell façade walls by means of cylindrical and conical surfaces.

The authors have also initiated work on parametric shaping of structural systems dedicated to the considered complete architectural free-forms and their structures. Specific structural systems stiffening the essential edges of the discussed free-forms are necessary because of the complex structure of the roof and façade forms as well as the varied inclination of their elements, such as the oblique edges of the eaves and façade.

In the authors' previous papers, a different approach to the creation of spatial structures composed of many individual free-forms roofed with structures composed of many individual transformed folded shells was presented. The main aim of the aforementioned papers was to use the wide possibilities offered by combining many reference tetrahedrons Γ_i (Figure 26), such as the ones used for the case of the presented example structure Σ presented in Figures 16–25.

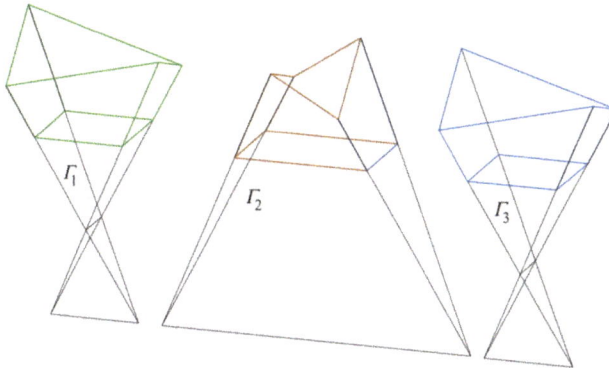

Figure 26. Structure exploded into three reference tetrahedrons Γ_i.

Different types of these reference tetrahedrons may be joined into one complex spatial reference network (Figure 27). Some of the types are presented in this article. The spatial network is used to give the curvature and overall dimensions of the complex roof shell of the architectural structure being sought. The spatial cells of such a network—named reference tetrahedrons—help in locating single free-forms of various specific dimensions of their roofs and façades.

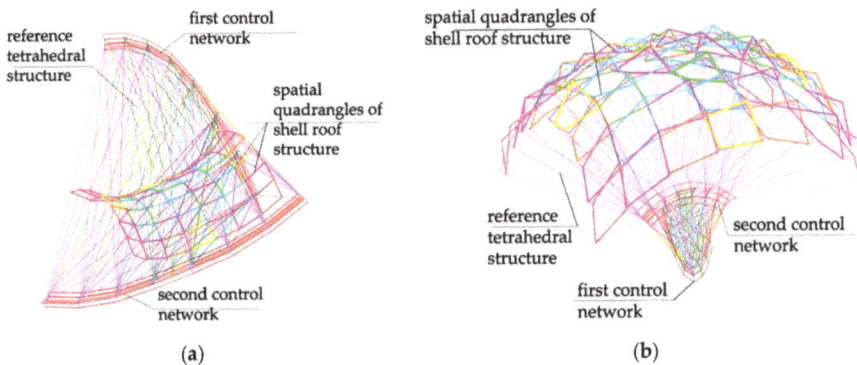

(a) (b)

Figure 27. Determined by vertices of two control networks, two roof shell structures of various kinds characterized by (**a**) the negative Gaussian curvature and (**b**) the positive Gaussian curvature.

Author Contributions: J.A. carried out research and analyses, visualized and interpreted the results, created models and method as well as wrote all sections of the paper. J.A. was the supervisor and the project administrator. A.P. participated in the concept and interpretation of the results as well as funding acquisition.

Funding: The resources of the Rzeszow University of Technology.

Conflicts of Interest: The authors declare no conflicts of interest.

References

1.   Abel, J.F.; Mungan, I. *Fifty Years of Progress for Shell and Spatial Structures*; International Association for Shell and Spatial Structures: Madrid, Spain, 2011.

2.   Foraboschi, P. The central role played by structural design in enabling the construction of buildings that advanced and revolutionized architecture. *Constr. Build. Mater.* **2016**, *114*, 956–976. [CrossRef]

3.   Foraboschi, P. Structural layout that takes full advantage of the capabilities and opportunities afforded by two-way RC floors, coupled with theselection of the best technique, to avoid serviceability failures. *Eng. Fail. Anal.* **2016**, *70*, 377–418. [CrossRef]

4.   Foraboschi, P. Modeling of collapse mechanisms of thin reinforced concrete shells. *J. Struct. Eng. ASCE* **1995**, *121*, 15–27. [CrossRef]

5.   Foraboschi, P. Optimal design of glass plates loaded transversally. *Mater. Des.* **2014**, *62*, 443–458. [CrossRef]

6.   Liu, Y.; Zwingmann, B. Carbon Fiber Reinforced Polymer for Cable Structures—A Review. *Polymers* **2015**, *7*, 2078–2099. [CrossRef]

7.   Sibley, M. Let There Be Light! Investigating Vernacular Daylighting in Moroccan Heritage Hammams for Rehabilitation, Benchmarking and Energy Saving. *Sustainability* **2018**, *10*, 3984. [CrossRef]

8.   Moavenzadeh, F. *Global Construction and the Environment Strategies and Opportunities*; John Wilej ans Sons Inc.: Hoboken, NJ, USA, 1994.

9.   Foraboschi, P. Versatility of steel in correcting construction deficiencies and in seismic retrofitting of RC buildings. *J. Build. Eng.* **2016**, *8*, 107–122. [CrossRef]

10.   MacGinley, T.J. *Steel Structures Practical Design Study*, 2nd ed.; E&FN Spon: London, UK; New York, NY, USA, 2002.

11.   Obrębski, J.B. Observations on Rational Designing of Space Structures. In Proceedings of the Symposium Montpellier Shell and Spatial Structures for Models to Realization IASS, Montpellier, France, 20–24 September 2004; pp. 24–25.

12.   Rębielak, J. Review of Some Structural Systems Developed Recently by help of Application of Numerical Models. In Proceedings of the XVIII International Conference on Lightweight Structures in Civil Engineering, Łódź, Poland, 7 December 2012; pp. 59–64.

13.   Gürlich, D.; Reber, A.; Biesinger, A.; Eicker, U. Daylight Performance of a Translucent Textile Membrane Roof with Thermal Insulation. *Buildings* **2018**, *8*, 118. [CrossRef]

14.   Yang, L.; Cui, L.; Li, Y.; An, C. Inspection and Reconstruction of Metal-Roof Deformation under Wind Pressure Based on Bend Sensors. *Sensors* **2017**, *17*, 1054. [CrossRef] [PubMed]

15.   Reichhart, A. *Geometrical and Structural Shaping Building Shells Made up of Transformed Flat Folded Sheets*; Rzeszow University of Technology: Rzeszów, Poland, 2002. (In Polish)

16.   Abramczyk, J. An Influence of Shapes of Flat Folded Sheets and Their Directrices on the Forms of the Building Covers Made up of These Sheets. Ph.D. Thesis, Rzeszow University of Technology, Rzeszów, Poland, 2011. (In Polish)

17.   Abramczyk, J. Principles of geometrical shaping effective shell structures forms. *JCEEA* **2014**, 5–21. [CrossRef]

18.   Abramczyk, J. *Shell Free Forms of Buildings Roofed with Transformed Corrugated Sheeting*; Rzeszow University of Technology: Rzeszów, Poland, 2017.

19.   Reichhart, A. Principles of designing shells of profiled steel sheets. In Proceedings of the X International Conference on Lightweight Structures in Civil Engineering, Rzeszow, Poland, 5–6 December 2004; pp. 138–145.

20.   Abramczyk, J. Shaping Innovative Forms of Buildings Roofed with Corrugated Hyperbolic Paraboloid Sheeting. *Procedia Eng.* **2016**, *161*, 60–66. [CrossRef]

21.   Abramczyk, J. Responsive Parametric Building Free Forms Determined by Their Elastically Transformed Steel Shell Roofs Sheeting. *Buildings* **2019**, *9*, 46.

22.   Abramczyk, J. Parametric shaping of consistent architectural forms for buildings roofed with corrugated shell sheeting. *J. Archit. Civil Eng. Environ.* **2017**, *10*, 5–18.

23.   Prokopska, A.; Abramczyk, J. Innovative systems of corrugated shells rationalizing the design and erection processes for free building forms. *J. Archit. Civil Eng. Environ.* **2017**, *10*, 29–40. [CrossRef]

24.   Reichhart, A. Corrugated Deformed Steel Sheets as Material for Shells. In Proceedings of the International Conference on Lightweight Structures in Civil Engineering, Warsaw, Poland, 26–29 December 1995.

25. Abramczyk, J. Integrated building forms covered with effectively transformed folded sheets. *Procedia Eng.* **2016**, *8*, 1545–1550. [CrossRef]

26. Medwadowski, S.J. Symposium on Shell and Spatial Structures: The Development of Form. *Bull. IASS* **1979**, *70*, 3–10.

27. Medwadowski, S.J. The interrelation between the theory and the form of shells. *Bull. IASS* **1979**, *70*, 41–61.

28. Saitoh, M. *Recent Spatial Structures in Japan*; J. JASS: Madrid, Spain, 2001.

29. Makowski, Z.S. *Steel Space Structures*; Michael Joseph: London, UK, 1965.

30. Makowski, Z.S. *Analysis, Design and Construction of Double-Layer Grids*; Applied Science Publishers: London, UK, 1981.

31. Chilton, J. *Space GridStructures*; Architectural Press: Oxford, UK, 2000.

32. Abramczyk, J. Building Structures Roofed with Multi-Segment Corrugated Hyperbolic Paraboloid Steel Shells. *J. Int. Assoc. Shell Spat. Struct.* **2016**, *2*, 121–132. [CrossRef]

33. Tahmasebinia, F.; Niemelä, M.; Sepasgozar, S.M.E.; Lai, T.Y.; Su, W.; Reddy, K.R.; Shirowzhan, S.; Sepasgozar, S.; Marroquin, F.A. Three-Dimensional Printing Using Recycled High-Density Polyethylene: Technological Challenges and Future Directions for Construction. *Buildings* **2018**, *8*, 165. [CrossRef]

34. Abramczyk, J. Shape transformations of folded sheets providing shell free forms for roofing. In Proceedings of the 11th Conference on Shell Structures Theory and Applications, Gdańsk, Poland, 11–13 October 2017; Pietraszkiewicz, W., Witkowski, W., Eds.; CRC Press Taylor and Francis Group: Boca Raton, FL, USA, 2017; pp. 409–412.

35. Bathe, K.J. *Finite Element Procedures*; Englewood Cliffs NJ Prentice Hall: Bergen County, NJ, USA, 1996.

36. Gergely, P.; Banavalkar, P.V.; Parker, J.E. The analysis and behavior of thin-steel hyperbolic paraboloid shells. In *A Research Project Sponsored by the America Iron and Steel Institute*; Report 338; Cornell University: Ithaca, NY, USA, 1971.

37. McDermott, J.F. Single layer corrugated steel sheet hypars. *Proc. ASCE J. Struct. Div.* **1968**, *94*, 1279–1294.

38. Egger, H.; Fischer, M.; Resinger, F. Hyperschale aus Profilblechen. Der Stahlbau, H., Ed.; Ernst&Son: Berlin/Brandenburg, Germany, 1971; Volume 12, pp. 353–361.

39. Davis, J.M.; Bryan, E.R. *Manual of Stressed Skin Diaphragm Design*; Wiley: Granada, London, 1982.

40. Prokopska, A. Creativity Method applied in Architectural Spatial cubic Form Case of the Ronchamp Chapel of Le Corbusier. *Syst. J. Transdiscipl. Syst. Sci. (JTSS)* **2007**, *12*, 49–57.

41. Prokopska, A. *Methodology of Architectural Design Preliminary Phases of the Architectural Process*; Publishing House of Rzeszow University of Technology: Rzeszów, Poland, 2018.

42. Prokopski, G.B.; Prokopska, A. Computer based Assisting The Preliminary (Preparatory) Phase of the Architectural Process. *JTSS* **2008**, *13*, 41–50.

buildings

MDPI

Article

Responsive Parametric Building Free Forms Determined by Their Elastically Transformed Steel Shell Roofs

Aleksandra Prokopska and Jacek Abramczyk *

Department of Architectural Design and Engineering Graphics, Rzeszow University of Technology,
Al. Powstańców Warszawy 12, 35-959 Rzeszów, Poland; aprok@prz.edu.pl
* Correspondence: jacabram@prz.edu.pl; Tel.: +48-795-486-426

Received: 8 January 2019; Accepted: 11 February 2019; Published: 14 February 2019

Abstract: The article concerns the unconventional architectural forms of buildings roofed with transformed shells made up of thin-walled steel fold sheets, and a parametric description of how they are shaped. Complicated deformations of flanges and webs, as well as the complex static–strength work of the folds in a shell roof, demand the creation of simplified models regarding the parameterization of such shells and their integration with the general forms of the buildings. To obtain favorable results, it was necessary to write computer applications because of both the complicated problems related to the significant limitations of the transformations, as well as the great possibilities of shaping shell roofs by means of directrices of almost free shape and mutual position. The developed procedures enable the prediction of shapes and states of all the folds in the designed shell. They take account of two basic conditions related to these restrictions, which guarantee that the folds encounter little resistance when matching their transformed forms to the roof directrices, and that their initial effort was as low as possible. The developed procedures required solving a number of issues in the fields of architecture, civil engineering, and structures, and are illustrated with an example of shaping one unconventional architectural form. The interdisciplinary study explains a new insight into shaping such forms.

Keywords: corrugated shell roof; free-form building; architectural form; folded sheet; thin-walled profile; shape transformation; steel construction

---

1. Introduction

Curved shells, whether stiffened or not with structural ribs, that carry dead and live loads have been a great challenge for the engineers and architects of every era. In subsequent epochs, not only were the materials, weight, static diagrams, stiffness of structural elements and joints, spans, and durability of the designed shells and entire buildings changed, but their visual [1] attractiveness, form coherence, and architectural sensitivity to the natural and built environments have been modified as well [2].

Since the Roman times, single-curvature shell vaults have been used more and more often, including especially barrel and cross vaults. Since the Gothic style, doubly-curved roof shells with a positive Gaussian curvature [2] have been built, which was a result of the expected compressive stresses in them [3]. Stiffening or supporting ribs have been used to join complete smooth shells into a composite shell structure [3].

The issues related to the search for thin-walled concrete shells transferring a characteristic load were presented by H. Isler inter alia in [4]. He created models based on the nature-based solutions and conducted experiments with surfaces similar to the so-called minimum surfaces.

Examples and procedures preventing the destruction of reinforced concrete shells were presented by Foraboschi in [5]. An additional factor that causes damages to roof shells is dynamic influences.

Foraboschi discussed the appropriate procedures to prevent unfavorable dynamic influences in [5,6]. In addition, in the areas affected by seismic influences, the roof shell structure should be designed to improve the durability and bearing capacity of the designed building [7].

Shell roofs can simultaneously perform various functions. The multidimensionality of the issues related to their design, construction, and maintenance requires a comprehensive, parametric approach to shaping diversified unconventional architectural forms and the structures of entire buildings roofed with the shells [8]. The aspects of the parametric description of the architectural forms are under consideration in this paper. In particular, this paper proposes a parameterization of such roofs that is made up of many nominally plane thin-walled folded steel sheets connected to each other along their longitudinal edges into single continuous plane strips. Subsequently, each strip is transformed into a corrugated shell roof (Figure 1) as a result of spreading the strip on two skew directrices passing transversally to the fold's directions [1].

Figure 1. A roof shell structure composed of many transformed fold shell strips: (**a**) view from the outside; (**b**) view from the inside.

One of the characteristics of the considered transformed shells is that the directrices stiffen their transverse edges, but their longitudinal edges must be stiffened with additional edge elements in order to maintain the straightness of the border folds in each strip (Figure 1b) [1]. This is the first limitation in shaping the transformed shells, which induces additional effort besides the initial stresses caused by the shape transformations that are determined by arranging and pressing each folded strip to the roof directrices. It should be noted that the technique and direction of the pressure of each fold to the directrices should result in the smallest cross-sectional change of the fold, so as not to unduly reduce its capacity and stability.

If the directrices are parallel to each other (Figure 2c) [9], the shape transformations do not result in significant values regarding the initial stresses, because the curvatures that are used in most building shell roofs and roof directrices are not unduly large, and the stresses need not be included in the static-strength calculations [1]. In this case, the shells can take the forms of various cylindrical surfaces. However, if the directrices are not parallel lines [8,9], the folds are twisted (Figure 2a) or twisted and bent (Figure 2b,d). Moreover, the deformations of their webs and flanges can be considerable and different both along the length of the same fold and in the adjacent folds in a shell. These differences may result in substantial values of compressive stresses in the fold's half-lengths and tensile stresses at both of the transverse ends of each twisted fold, depending on the degree of the fold's twist.

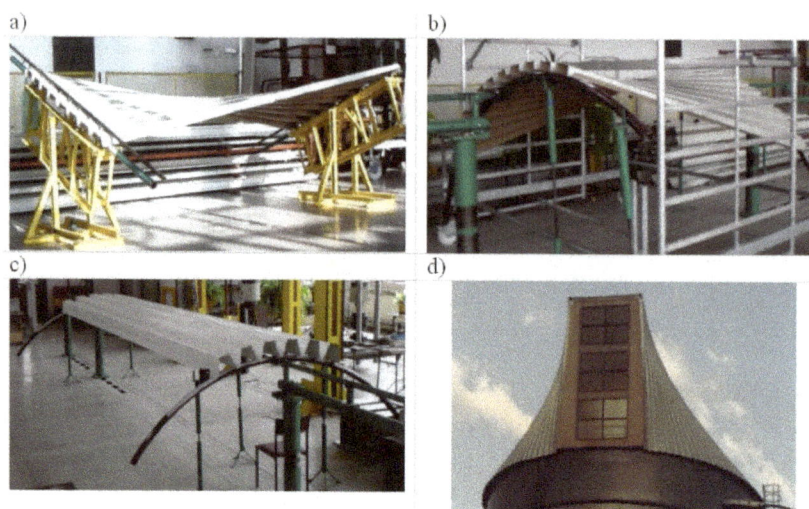

Figure 2. Corrugated shell sheets spread on: (**a**) straight directrices; (**b,c**) curved directrices; and (**d**) straight and curved directrices.

The longitudinal edges of each twisted or twisted and bent fold, similar to the longitudinal axes of each pair of adjacent folds in the shell, are skew straight lines, which result in different cross-sections of each shell fold along its length. The experimental tests and computer analyses carried out by Reichhart and Abramczyk [1,8] showed that each such transformed shell fold works effectively when its contraction occurs halfway along its length. In this case, the tensile stresses appearing at both transverse ends of the fold are comparable, and they balance the compressive stresses appearing in the middle part of the fold along the length.

Moreover, the distribution of the above-mentioned stresses in the fold's flanges and webs shows that each such transformed fold tends to bend its longitudinal edges with the convexity halfway along the length of the longitudinal edges directed to the outside, thereby affecting the adjacent folds in a shell (Figure 3). The action of the fold has to be balanced by the forces affecting the fold and coming from its neighboring folds. Transformed folds are designed to carry their own weight as well as the characteristic load, so the initial effort resulting from the shape transformations has to be limited appropriately.

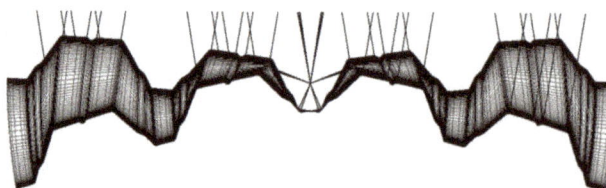

Figure 3. An exact computational folded model of a nominally plane folded sheet transformed into a shell shape.

These influences of adjacent folds in shells have not yet been researched well enough, and the descriptions presented in the available literature are too general. However, the results of the experimental tests and computer analyses [1,10] indicate a large variety of possible unconventional forms of thin-walled folded sheeting transformed from flat to spatial forms, despite these initial stresses. The variety results from the great freedom in the adoption of the shapes and the mutual position of roof directrices, as well as the location of the fold directions in relation to the directions of directrices. The non-perpendicularity of the directions of folds and directrices results in an oblique cutting of both transverse ends of the shell folds [8]. A parametric description of the relationships between these supporting conditions of shell folds and the shapes of these folds allows the use of computer programming technology to create simplified smooth models of these folds and entire shells for engineering developments. The supporting conditions depend on the shapes and mutual position of the roof directrices, and are called boundary conditions. For scientific purposes, that is, for an incremental non-linear dynamic analysis of the static and strength properties of the shell folds as structural elements (Figure 4) [10], authors use advanced programs such as ADINA, for example [11].

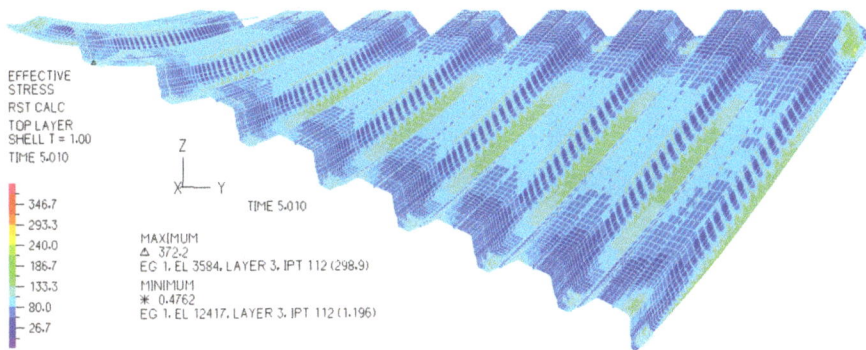

Figure 4. An exact computational folded model of a nominally plane folded sheet transformed into a shell shape.

In order to understand the parametric description presented in the present article, some problems should be explained. For engineering developments, regular geometrical surfaces (Figure 5) are employed to model subsequent shell folds in each transformed shell roof. It is possible to find only one shell shape of a transformed fold, which is assigned to the calculated border conditions resulting from the geometrical supporting conditions in the shell, such that its transformation is effective [1,8]. The characteristic of this shape is that its contraction passes halfway along its length transversally to the fold's longitudinal axes. Figure 5 shows the contraction line, which is called the line of striction, and is denoted as s. As a result, the effectively transformed fold can be spread on the roof directrices relatively freely, that is, with the lowest possible pressing forces. Furthermore, its impact on the forms of adjacent folds in the shell is the least possible. In this way, the effort is optimized to the lowest possible level. The above-mentioned pressing forces are needed to fix the fold's ends to the roof directrices.

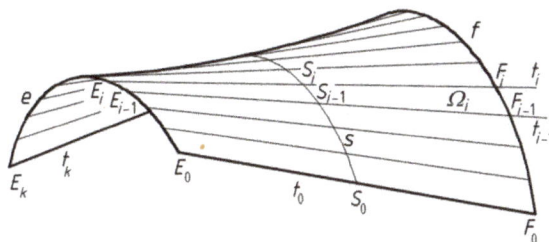

Figure 5. Simplified smooth shell model of a transformed shell roof with the s line of striction and rulings t_i modeling the longitudinal borders of subsequent folds in a shell.

For the effective fold transformations, interdependence between the geometrical supporting conditions and the obtained shell forms of a transformed fold can be used. In these cases, the freedom of the transverse width and height increments of each shell fold forming the transformed sheeting is ensured, and the various attractive and innovative shapes of shell roofs and contraction curves of relatively big curvatures can be achieved (Figure 6) [1,9]. If the fold does not have the freedom of the transverse width increments due to the strong stiffening of its longitudinal edges shared with its adjacent folds, or if the assembly technique causes additional forces varying the effective widths of the fold ends and their supporting lines, the aforementioned interdependence cannot be used.

(a)　　　　　　　　　　　　　　　　　　(b)

Figure 6. A transformed roof shell whose inside bottom layer is folded sheeting: (**a**) view from the inside; (**b**) view from the outside.

The application of the well-known conventional design methods [12,13], which is known from the traditional courses of theory of structures, in the shaping of transformed shell roof forms is rather ineffective, because it usually results in high values of normal and shear stresses, local buckling, and the distortion of thin-walled flanges and webs of shell folds. It is often impossible to assemble the designed shell sheeting into skew roof directrices because of the plasticity of the fold's edges between flanges and webs. Reichhart developed various methods for calculating this arrangement and the length of the supporting lines of all the folds in the transformed sheeting [1]. Abramczyk improved the method [8,9] so that the transformation would cause the smallest possible initial stresses of the shell folds.

Therefore, the designer may have to face, and cope with, some problems that arise from using unconventional methods for shaping the general architectural forms of buildings roofed with transformed folded steel sheets, and striving for the relatively simple implementation of the designed innovative forms. The solution of these problems is the priority. The main task is to achieve the geometrical, architectural, and structural cohesion of all the elements of each free-form building, and its shell roof in particular [14]. This aim is accomplished by creating a parametric description of such building free forms and, in the near future, their specific structural systems [15,16] based on the geometrical and mechanical characteristics of the transformed sheeting [10].

Other difficult issues related to the shape transformation of thin-walled folded shells are the diversified supporting conditions, which are calculated for subsequent folds in the same shell. The diversification results from the mutual skew position of the roof directrices, which results in different twist degrees for the subsequent folds in the shell. Thus, the twist degree is the basic border condition that is calculated for each fold in a roof shell, and affects the shell shape of the fold. The aforementioned interdependence between the supporting conditions and the shape of each shell fold is reduced to the interdependence between its twist angle degree and supporting line length. The lengths of the supporting lines of all the subsequent folds along each directrix have basic significance when searching for the fold's shell shapes.

An additional complication is caused by none of the directrices in relation to the transformed shell sheeting needing to be symmetrical nor congruent. This means that the sum of the calculated lengths of the supporting lines of all the subsequent shell folds may be different from the length of both employed directrices, so one of the directrices does not have to be completely covered with the sheeting. The differences can reach more than one meter. In this case, changing at least one of the parameters used in the presented description allows the shape and length of one of the directrices to be adjusted to the width of the whole roof shell. Attempts to change the widths of the fold's transverse ends during their assembly in the shell roof without recalculations are unjustified, because they cause an unnecessary increase in the initial stresses and most often need high forces that can even result in the plasticity of the fold's flanges and webs. This increase may also be a result of miscalculations related to the optimal fold's transformed shapes.

For engineering developments, each shell fold can be modeled with a simplified smooth sector of a warped surface [17,18]. The sum of all such sectors is a continuous edge structure. On the basis of this structure, one single smooth shell sector approximating this structure and modeling the entire transformed roof shell is created. In this case, the loft function of many graphics computer programs can be used.

2. Critical Analysis

The two straight lines shown in Figure 7, x and y, are two rulings of a specific kind of hyperbolic paraboloids. This type of hyperbolic paraboloids, which is characterized by such lines being perpendicular to each other, is often used to model the transformed corrugated shells, which are called hypars. Figure 7a shows a simplified, smooth model of a transformed corrugated shell. The shell has quite unique general geometrical properties that are similar to the central sector of the hyperbolic paraboloid symmetric about the x and y axes of the Euclidean coordinate system $[x, y, z]$. These axes belong to two various families of rulings of the paraboloid, and divide this paraboloid into four congruent units, which are designated as one, two, three, and four. The dimensions of this central sector are represented by the capital letters A, B, and C. These dimensions are the absolute values of the coordinates of four vertices belonging to the edge line of the central section in $[x, y, z]$.

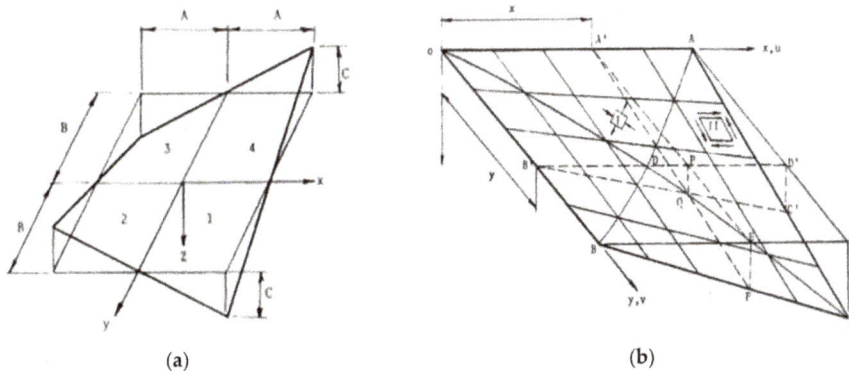

(a) (b)

Figure 7. Basic hypar units: (**a**) a central sector of hyperbolic paraboloid; (**b**) a complete unit: one-fourth of the central sector.

Such surfaces were employed by, for example, McDermott et al. [13,19]. Corrugated shell roofs can be shaped as central sectors, or one-fourth of the central sectors of hyperbolic paraboloids (Figures 7 and 8). Various configurations of shell structures composed of such sectors were used by Fisher et al. [20,21]. The diversity of shell roof forms may be slightly improved by using the computer program developed by Gioncu and Petcu [22].

Figure 8. (**a**) The experimental hyperbolic paraboloid shell; (**b**) the erected hyperbolic paraboloid shell structure.

The open thin-walled profile and orthotropic properties of the transformed sheeting result in many advantages and disadvantages of the discussed architectural free forms. This makes it necessary for the shell shape of each fold in the shell sheeting to be optimized in relation to the supporting conditions to obtain the lowest possible negative stresses and strains, as well as attractive shell shapes [8,12].

Based on his experimental tests [1], Reichhart developed an innovative simple method for shaping free-form roofs made up of elastically transformed folded sheets. The concept of his method lies in using the geometrical and mechanical properties of nominally plane-folded steel sheets transformed into rational shell roofs [23]. He employed some basic characteristics of such transformed sheeting observed during his experimental tests. Reichhart's experimental sheets were supported by straight directrices [24]. Kiełbasa created a computational folded model of freely twisted sheets using Reichhart's concept [25].

Abramczyk found Reichhart's concept a very rational approach [8]. However, he has proved that the simplifications made by Reichhart cause very significant errors in roof shell shaping, because they lead to ineffective forced shape transformations and induce unnecessarily high stresses and even plastic deformations of the shell fold's walls.

Computer programming enables the search for innovative diversified corrugated shell roofs and entire building forms [16]. Taking advantage of this possibility, Abramczyk developed a method for the intuitive shaping of free-form buildings covered with plane glass elevations and transformed shell steel roofs [14], and the creation of their simplified models. His method is constantly evolving [26,27], used by graduates [28,29], and extended to complex free-form structures [30] (Figure 9). Some of Abramczyk's tests and analyses were carried out on his computational folded models [10] created in a numerical program called Advances in Dynamic Incremental Nonlinear Analyses [11].

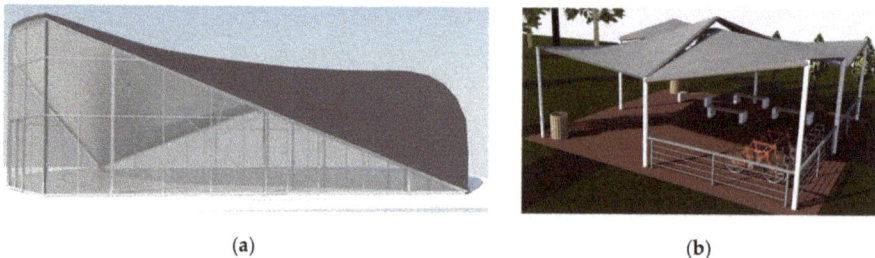

Figure 9. Architectural phase of: (**a**) a complete free-form building, (**b**) a shell roof structure composed of four hyperbolic paraboloids.

3. Aim and Scope

The aim of the paper is to carry out and present a novel parametric description of the process of shaping the following:

1. Shell roof forms made up of nominally flat thin-walled folded sheets transformed into spatial forms as a result of connecting their longitudinal edges to obtain a single corrugated strip and arranging the strips on two directrices,
2. General architectural forms of whole buildings using the above transformed sheeting to obtain attractive shapes, dimensions, proportions, and slopes of all the complete building elements such as façades, eaves, and roofs, and their characteristic lines, planes, and surfaces. The goal of this process is to achieve an internal shape integration of the shaped architectural form.

The proposed parametric description allows two things. Namely, complex solutions of the problems presented in the previous sections can be simplified, and, second, a technically useful algorithm that is implemented in computer programs and developed by one of the authors can be created. A method based on the above description and supported by the aforementioned computer programs assists the designer in the process of searching for the expected diversified architectural free forms.

In the article, the authors assume that the proposed parametric description must be presented for an example regarding the search for a visually appealing and internally consistent architectural form characterized by a relatively free general form roofed with a transformed roof sensitive to harmonious incorporation into the expected built environment. For this purpose, it was decided that the object of the search is the simplest set of parameters chosen from all the parameters employed in the presented parametric description. Moreover, the values can be assigned to the selected parameters so that the internal integrity and external sensitivity of the searched form is achieved. Additionally, it was assumed that:

1. The directrices of the roof shell are curved,
2. The curvatures of the transformed shell have to be big enough to achieve the appropriate dimensions of its parts, which should be visible from the directions parallel to the horizontal building base plane and constitute a significant part of the architectural form that is sought.

4. Concept

In order to achieve the objectives proposed in the previous section, the following concept of activities is adopted. In the first step, a parametric description of the considered general building forms is used. This step results in a simplified, flat-walled model Σ consisting of four quadrangles that have vertices at P_i, B_i ($i = 1$ to 4) and represent the four elevation walls of a building (Figure 10). Both of the Σ forms shown in Figure 10 have to be built on the basis of the earlier created reference tetrahedrons Γ, which are defined by the means of four adopted vertices, H_i. One edge of each of these quadrangles is also a segment of a spatial closed line $B_1B_2B_3B_4$, which is a model of straight roof eaves. When the roof directrices e and f are curved, they are usually adopted in planes γ_1 and γ_3 (or γ_2 and γ_4) of the opposite elevation walls, as shown Figure 10b. It is often assumed that two sides of line $B_1B_2B_3B_4$ are the chords of the adopted arcs e and f.

In the second step of the algorithm, a parametric description of the smooth shell models of the building's roofs is used. The accuracy of the models is satisfactory for engineering developments.

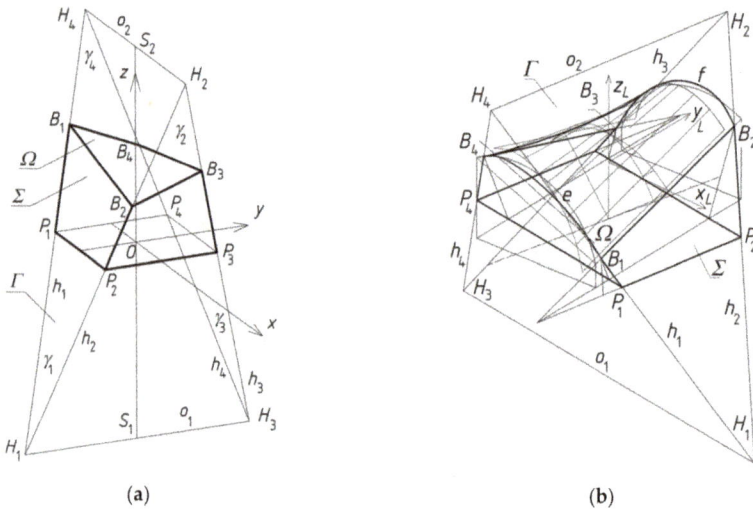

Figure 10. Free-form shaping using: (**a**) straight directrices; (**b**) curved directrices *e* and *f*.

In the third step of the algorithm, a parametric description of the basic elements of the building such as flat-walled elevations and shell roofs is used. It includes the thickness of elevation walls and roof shells, the division of each elevation wall into areas creating regular patterns, and the protrusion of roof eaves outside the outline of the elevations.

In the final step of the algorithm, a parametric description of the building structural system dedicated to the folded shell roof and oblique flat-walled glass elevations is used. The description of this step goes beyond the scope of the present paper. It is also possible to extend the method to structures composed of several individual free forms that share walls, such as the ones presented in Figure 11.

Figure 11. Various types of free-form structures. (**a**) Configuration 1; (**b**) Configuration 2.

5. Parametric General Building Free Forms

Four flat quadrangles Σ and one sector Ω of a warped surface, which are shown in Figure 12, are the basic objects that are built in this step of the algorithm. They create a simplified model representing the general form of a free-form building. When its roof directrices are two curves, they are the lines limiting two of the aforementioned quadrangles modeling two opposite elevation walls, as shown in Figure 12b. These forms belong to the second basic kind of the architectural free forms discussed in the paper.

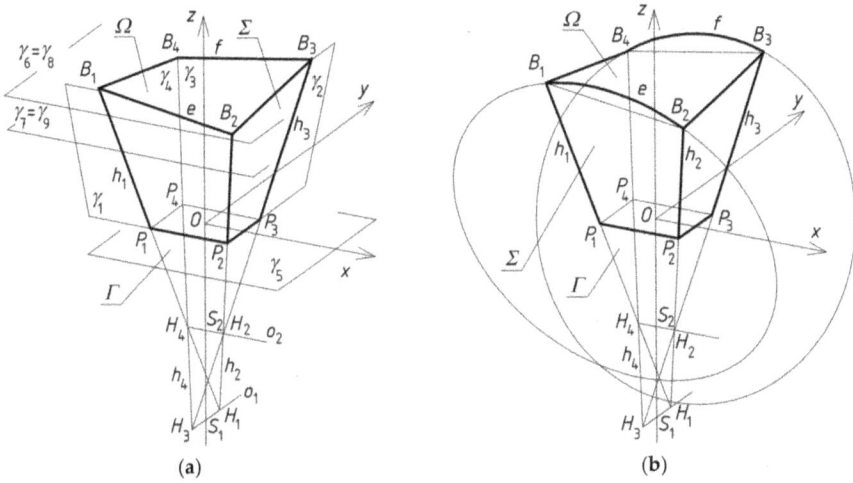

Figure 12. Two simplified models of a free-form building roofed with transformed shells supported by: (a) straight directrices; and (b) curved directrices.

Nine planes γ_i (i = one to nine) are the auxiliary objects in modeling the general building form. The first four planes (for i = one to four) allow the construction of the aforementioned quadrangles $P_i P_{i-1} B_{i-1} B_i$ as follows. Each two adjacent quadrangles have a common edge contained in the straight line h_i (i = one to four), which is the intersection of two adjacent planes γ_{i-1} and γ_i. For example, the $P_1 P_2 B_2 B_1$ and $P_2 P_3 B_3 B_2$ quadrangles, which are shown in Figure 12, have the common edge $P_2 B_2$ contained in h_2. These four planes γ_i (i = one to four) define the so-called reference tetrahedron Γ. The plane γ_5 contains the building's horizontal base $P_1 P_2 P_3 P_4$. The planes γ_j (j = six to nine) define the levels of the corners B_i (i = one to four) of the building eaves. These points belong to h_i, too. The opposite planes γ_i and γ_{i+2} intersect in the axes o_1 or o_2 of Γ. The neighboring edges h_i intersect at the vertices H_i of the reference figure Γ.

The activities carried out on the aforementioned facilities include:

- adopting an orthogonal coordinate system $[x, y, z]$ in three-dimensional space,
- accepting any two points S_1 and S_2 on the axis z,
- passing axis o_1 || y through point S_1,
- passing axis o_2 || x through point S_2,
- selecting vertices H_i (i = one to four) on axes o_1 and o_2,
- defining each straight line h_i by means of vertices H_i,
- obtaining planes γ_i (i = one to four) defined by the respective pairs of neighboring h_i,
- creating corners P_i of the rectangular building base as the points of the intersection of plane γ_5 with each h_i,
- determining all the corners of the roof eaves, B_i, as the points of the intersection of planes γ_j (j = six to nine) with h_i, and
- defining directrices e and f contained in any two opposite planes γ_1 and γ_3 or γ_2 and γ_4.

The following parameters describing the building general forms were adopted in the algorithm:

1. ptr_k (for k = one to six) representing the lengths of the sections: $S_2 O$, $S_1 O$, $S_1 H_1$, $S_1 H_3$, $S_2 H_2$, and $S_2 H_4$,
2. ptr_r (for r = seven to 10) representing the distances of planes γ_j (j = six to nine) from plane $\gamma_5(x, y)$ of the building base,

3. ptr_{10} representing the same ridges of roof directrices e and f, which are usually shaped in the form of circle arcs.

It is assumed that the architectural form example belonging to the last—that is, third—basic type (Figure 13), which is discussed below, will be determined on the basis of the parametric description proposed above. The values of the parameters adopted for shaping the sought architectural form are given in Table 1. From the engineering point of view, what is more important is the coherence of the parametric description of architectural forms and the obtained set of proportions between the parameters, rather than the individual values of these parameters. In order to present these proportions, it is assumed that the reference parameter is the width ptr_{13} of the general form along its base. The values of the proportions that are considered important and shown in Table 2 refer to this reference parameter.

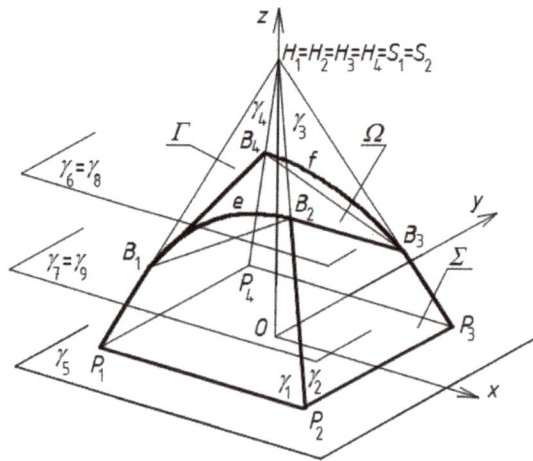

Figure 13. Simplified model of a building free form of the third type roofed with a transformed shell.

Table 1. Parameters adopted for the examined architectural free form from Figure 13.

Parameter	Value
$ptr_1 = ptr_2$	23,333.3
ptr_i for i = 3 to 6	0.0
$ptr_7 = ptr_9$	12,746.2
$ptr_8 = ptr_{10}$	12,746.2
ptr_{11}	1790.9
ptr_{12}	726.7
ptr_{13}	20,000.0

[1] values in millimeters.

Table 2. Proportions calculated for the examined architectural free form from Figure 13.

Proportion	Value
$ptr_1/ptr_{13} = ptr_2/ptr_{13}$	1.17
ptr_i/ptr_{13} for i = 3 to 6	0.0
$ptr_7/ptr_{13} = ptr_9/ptr_{13}$	0.64
$ptr_8/ptr_{13} = ptr_{10}/ptr_{13}$	0.64
ptr_{11}/ptr_{13}	0.090
ptr_{12}/ptr_{13}	0.036

In order to explain the attractiveness of a general form Σ, it is necessary to define additional parameters describing, for example, the width and height of the created architectural form, its roof and façade, the slopes of the roof eaves, edges, and the planes of façade walls. Such an action requires extensive considerations, and does not fall within the scope of the article. These issues are initially discussed in [13], and are related to a specific method, which is not presented in this article.

It is possible to create other sets of parameters defining the general building forms, as in the examples proposed by Abramczyk in [8,26]. However, his method is significantly more complex and requires a good spatial reasoning from the designer.

6. Parametric Shell Roofs

6.1. Introduction

Each regular warped surface [8,31] has straight rulings t_i and a line of striction s that intersects all of the rulings at the so-called central points S_i. When two rulings t_{i-1} and t_i of the surface approach each other at an infinitely short distance, then point S_i is the nearest point of t_i relative to ruling t_{i-1}. Line s is simply a contraction of the warped surface (Figure 14).

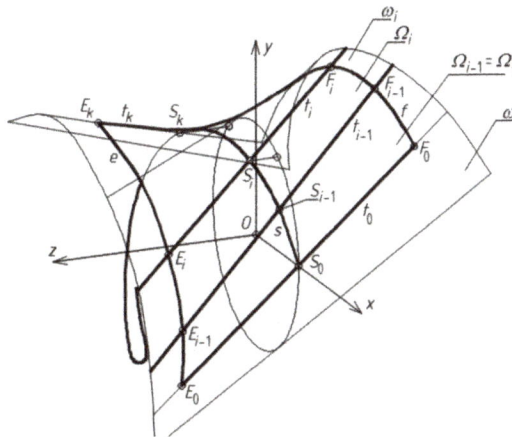

Figure 14. Sector Ω of warped surface ω modeling effectively transformed shell roof by means of contraction helix s $(S_0, \ldots, S_i, \ldots, S_k)$ of Ω.

It assumed that the created simplified models of all the discussed shell folds (possibly the entire single sheets if the curvatures of the considered shells are small) transformed into shell shapes are the central sectors Ω_i of different warped surfaces. In each such sector, there is a striction line s passing transversally, halfway along its length. Each sector is limited by rulings t_{i-1} and t_i lying in a proper distance from each other; the edge line of Ω_i is composed of the $E_{i-1}E_i$ section of directrix e, the E_iF_i section of ruling t_i, the F_iF_{i-1} section of directrix f, and the $E_{i-1}F_{i-1}$ section of ruling t_{i-1}.

Despite the attempts [8], it is impossible to invent one general mathematical equation defining all of the types of the warped surfaces used in modeling the discussed transformed roof shells. Therefore, all of the rulings t_i have to be determined in an approximate way on the basis of the adopted directrices e, f, and ruling t_{i-1}, which is calculated either at the previous step or adopted as t_0 at the beginning. The procedure of the latter solution is as follows. It is assumed that the E_{i-1}, F_{i-1}, and R_{i-1} points were constructed in the previous step, or $E_{i-1} = E_0$, $F_{i-1} = F_0$, and $R_{i-1} = R_0$ were calculated in the first step (Figure 15).

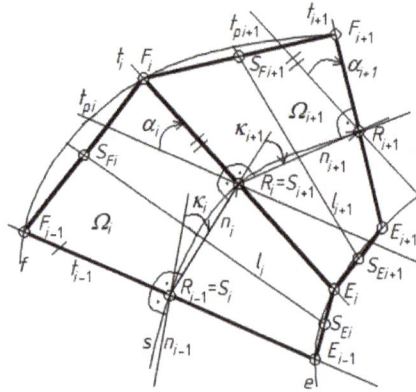

Figure 15. Shell sectors Ω_i and Ω_{i+1} modeling two subsequent effectively transformed shell roof folds limited by pairs of skew straight lines $\{t_{i-1}, t_i\}$ and $\{t_i, t_{i+1}\}$, which are determined on the basis of the lines n_i normal to the rulings t_i and t_{i-1}.

On the basis of these points, the positions of points E_i and F_i on directrices e and f and ruling t_i (E_i, F_i) are determined so that the straight line n_i perpendicular to t_{i-1} and t_i intersects line t_{i-1} at point R_{i-1}, and the area of the sector Ω_i modeling a shell fold after transformation is equal to the surface area of the same fold before transformation. The last condition can be used with satisfactory accuracy for engineering developments, but not for scientific research [8,10].

The positions of points E_i and F_i and ruling t_i are determined by the surface area and the twist angle of each transformed fold. The twist angle α of fold Ω_i is the angle of inclination of two planes: (S_{Ei}, S_{Fi}, E_i) and (S_{Ei}, S_{Fi}, F_i). The unit twist angle α_j of the fold is expressed as the quotient of the above twist angle α by length $|S_{Ei}S_{Fi}|$ of the fold. The degree of the fold's twist is the measure of the unit twist angle α_j, which is regarded as constant at the length of the fold. The length of the fold is taken as the length of the $S_{Ei}S_{Fi}$ section, where S_{Ei} and S_{Fi} are the midpoints of the $E_{i-1}E_i$ and $F_{i-1}F_i$ sections. Particular attention should be paid to the variation in the length and twist angle of the subsequent folds in a shell. The unit twist angle represents the basic geometrical supporting condition of the fold. Variable fold lengths indicate that the transverse ends of these folds have to be cut differently to adapt these ends to the directrices' directions.

Figure 15 shows the method of the shell fold's modeling by means of a special type of warped surfaces, i.e., such surfaces whose rulings are perpendicular to line s of striction. In a general case of a warped surface, its rulings are not perpendicular to its line of striction. Consequently, the subsequent straight lines n_i and n_{i+1} do not intersect the ruling t_i at the same point, which results in displacing points S_{i+1} and R_i along ruling t_i (Figure 16).

In this case, the procedure of searching for a simplified smooth model Ω_i of an effectively transformed shell fold should be extended by a second condition, in addition to the condition concerning the equality of surface areas of the fold before and after its transformation. This condition concerns the minimization of the length of the R_iS_{i+1} segment and positioning points S_{i+1}, R_i as close as possible to the midpoint of E_iF_i by looking for appropriate proportions between the lengths of the $E_{i-1}E_i$ and $F_{i-1}F_i$ lines. Consequently, the parametric description of shaping all of the shell folds has to cover both of the basic conditions that determine the efficiency of transformations of all the modeled shell folds in the transformed roof.

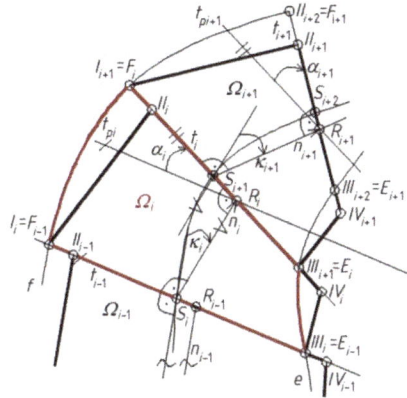

Figure 16. Models of two effectively twisted, bent, and sheared subsequent shell folds limited by pairs of two skew straight lines $\{t_i, t_{i+1}\}$, which are determined on the basis of a discontinuous sum of many $R_i S_i$ straight sections.

6.2. Simplified Smooth Parametric Models for Corrugated Roof Shells

The two conditions presented in the above introduction are implemented in the Rhino/Grasshopper program, which is orientated to the parametric modeling of geometric objects (Figure 17) and written by one of the authors. All of the individual objects and operations on these objects that are performed following the algorithm are defined by means of the flat rectangular graphic elements, which are named components, and are positioned on the named canvas of the Grasshopper background. The relationships between these objects are described by means of lines called connectors or wires. The application assists with creating inseparable simplified smooth shell models of all the subsequent shell folds (Figure 18). On the basis of the edge sum, a single smooth surface modeling of an entire building shell roof can be built.

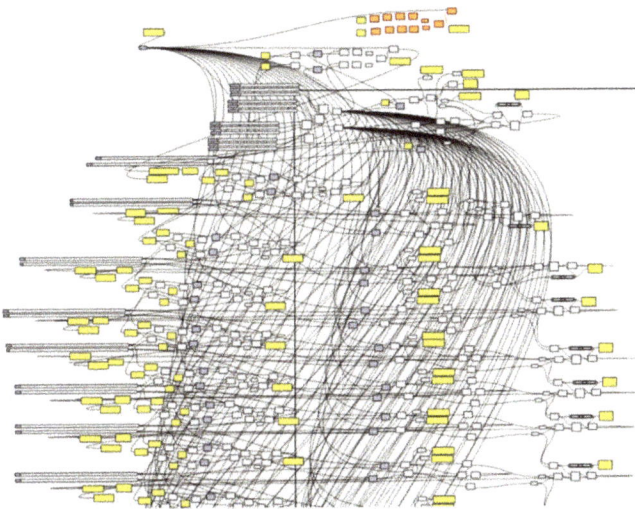

Figure 17. Part of the scheme of many objects creating the parametric algorithm implementing the authors' parametric description.

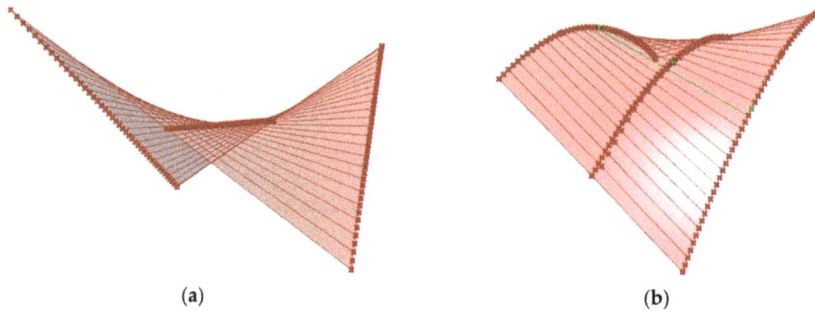

(a)   (b)

Figure 18. Simplified, smooth models of two transformed shells built in the Rhino/Grasshopper program and defined with: (**a**) straight directrices; and (**b**) curved directrices.

This step of the method's algorithm results in a smooth shell model of transformed roof sheeting. It is a sector of a warped surface, and can be formed as one of the two shapes presented in Figure 18. The geometrical properties of these smooth sectors represent the complex spatial deformed forms of all the folds of the transformed sheeting in a simplified way [23]. The current step ensures that the geometrical characteristics of these elements take account of the geometrical and structural properties of thin-walled folded sheets transformed in experimental tests, and their accurate models used in computer simulations carried out in the ADINA program [10].

The authors' application for the Rhino/Grasshopper program makes it possible to define two directrices e and f as algebraic lines, using two tetrads of the adopted points belonging to these directrices, whose coordinates are the entered initial data. If these directrices are straight, the coordinates of both their ends may only be entered. In this case, the four triads of sliders that are needed (Figure 19) to enter the coordinates of the ends of e and f are determined. Two components generating these straight directrices are also shown in Figure 19.

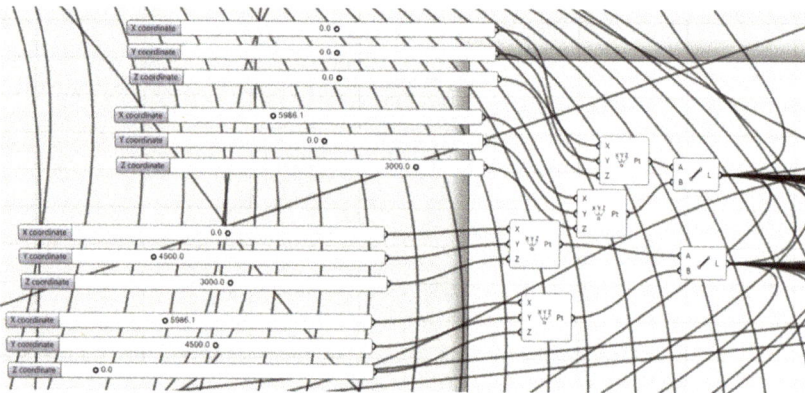

Figure 19. Four triad of sliders that allow three coordinates of the ends of two directrices, e and f, to be entered.

The first three sliders shown at the top of Figure 19 define three coordinates of the starting point of the first directrix f. The second three sliders allow entering the coordinate of the end point of f. The two next triads of sliders define the coordinates of the starting and end points of the second directrix, e.

Segments e_i and f_i of directrices e and f are the auxiliary objects of this step (Figure 20). They correspond to the supporting lines of the subsequent folds of a shell roof. The lengths of these segments are calculated, while the simplified shell model of each roof shell fold is developed. A central sector Ω_i of a warped surface limited by two rulings t_{i-1} and t_i and modeling this shell fold is sought by the means of points E_i and F_i displaced on e and f by means of the next two sliders, which are presented in Figure 21.

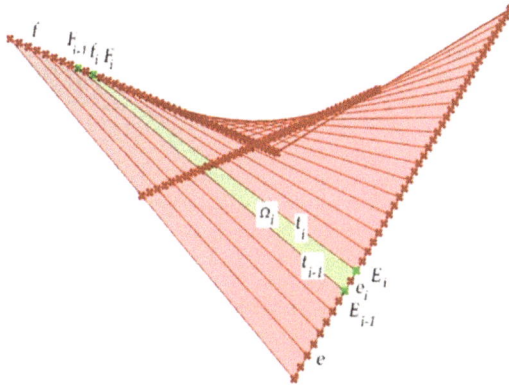

Figure 20. Narrow smooth shell sector Ω_i modeling a complete shell fold, which is created as the loft component and limited by two rulings, $t_{i-1}(E_{i-1}, F_{i-1})$ and $t_i(E_i, F_i)$, as well as two curves, e_i and f_i.

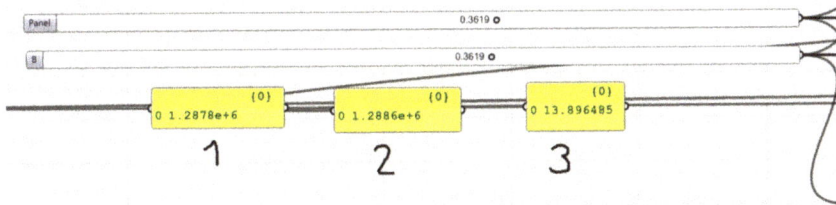

Figure 21. Two sliders and three panels assisting in meeting two conditions for effective fold transformations.

In the left panel signed '1', the information about the surface area (in millimetres) of the simplified model of the investigated fold before transformation is provided. In the middle panel signed '2', information on the surface area of the simplified model of this fold after transformation is given. The right-hand panel named '3' provides information about the distance between the contraction lines of the simplified models of two adjacent folds after transformation, which is the length of segment R_i and S_{i+1}, which was discussed in the introduction (Figure 16). The optimization of the shell shape of each designed fold in the shell consists of a change in the values of the above sliders such that the two numbers shown in panels '1' and '2' are equal to each other with the accuracy of about 10^4 mm, while the number in panel '3' should be close to zero, with the accuracy of up to 50 mm.

The two sliders presented in Figure 21 control the position of E_i on e and F_i on f, so they decide of the shape of Ω_i. The change of the values of these sliders, which causes a change of the values in panels '1' to '3', results in satisfying the two main conditions listed in the introduction, and is related to the surface areas of the simplified models of transformed folds and the optimal position of the contraction lines along the length of the folds. The conditions determine whether the modeled fold is effectively transformed or not.

Following the algorithm developed by the authors, the simplified model of each shell fold has to meet two conditions determining the harmonious and optimal work of all the folds in a shell. As we know, one concerns the strictly defined surface area of the smooth shell model of each shell fold in relation to its geometrical supporting conditions. It is represented in the authors' program by the green container shown on the left in Figure 22. The other condition concerns the contraction of each transformed corrugated shell. This condition requires that the contraction line passes transversely in relation to the directions of the shell folds through the middles of these folds along their lengths. The contraction is designated as s in Figures 14 and 15. In the authors' application of the Rhino/Grasshopper program, this condition is represented by the green container shown on the right in Figure 22. This container makes it possible to find rulings t_{i-1} and t_i perpendicular to contraction line s, and the distance between the R_i and S_i points that were discussed in the introduction.

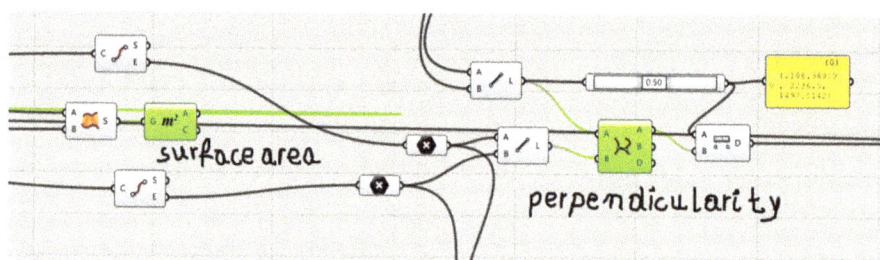

Figure 22. Two basic green components representing two basic conditions related to the fold's surface areas and line of contraction.

The two lines E_0E_i and F_0F_i shown in Figure 14 and called subcurves, and are used in the construction of E_i and F_i points on e and f. They are generated with the components shown in green in Figure 23. The application determines the e_i and f_i edge lines of Ω_i as differences between the subcurves E_0E_i and E_0E_{i-1}, and F_0F_i and F_0F_{i-1}, which were calculated for the adjacent folds in the shaped shell.

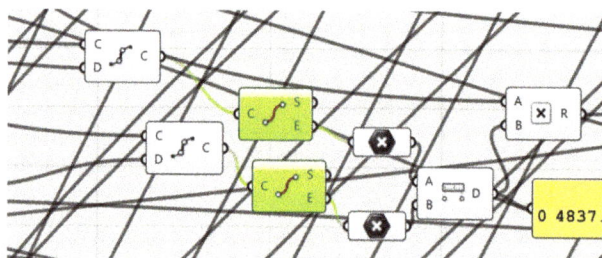

Figure 23. Components representing subcurves that are helpful in determining models e_i and f_i of the fold's supporting lines.

As a result of the comparison of the fold's surface areas before and after the shape transformation, the b_t width of each transformed fold along the appropriate directrix is displayed by means of the monitor component shown in Figure 24 as the number 289.944345, which was measured in millimetres. The nominal width of this fold before transformation was 280 mm. It is also displayed on one of the panels shown in Figure 24. The unit twist angle was equal to 8.2111045.

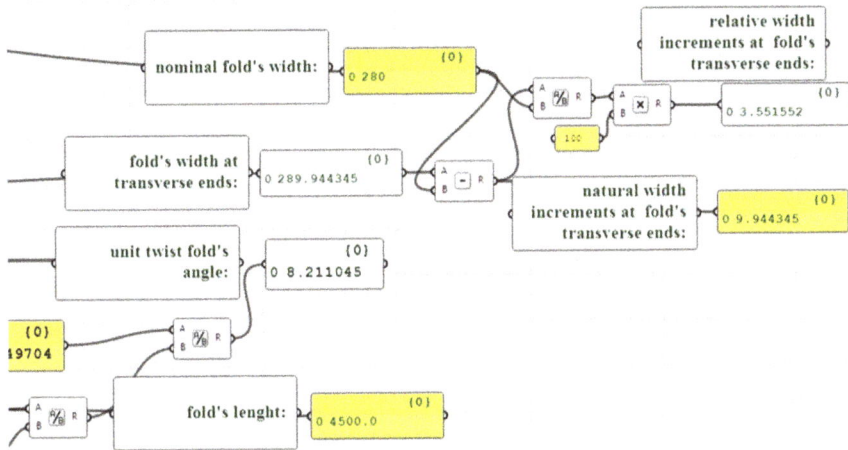

Figure 24. Important features of a simplified smooth shell model of the central fold of a shell, whose unit twisting angle α_j is equal to 8.211045°.

For engineering developments, the unit twist angle is the only significant parameter determining the supporting conditions of each transformed fold in a shell. The relative width increments of the shell fold along each directrix are the parameters that decide the shell form of the fold. For the shell fold, whose geometrical properties and supporting conditions are presented in the cells shown in Figure 21, the relative width increments along the both directrices are equal to 3.551552%. The relative increase b_{wr} in the fold's width is the quotient of the absolute width increment b_w of the transformed fold to the width b_0 of the fold before the transformation. It can be calculated following the formula:

$$b_{wr} = \frac{b_w}{b_0} \cdot 100\%, \tag{1}$$

In Figure 24, the b_w value is equal to 9.944345, which is the subtraction of $b_0 = 280$ mm from $b_t = 289.944345$. The b_w absolute width increment of the fold is then the difference between the width b_t of the fold after the transformation, and the b_0 value before the transformation. The width b_t of each fold in the shell is calculated as the length of the segment e_i of e or f_i of f (Figure 20).

Most of the relationships that were obtained during experimental tests are non-linear because of the big mutual displacements of adjacent folds, and the considerable deformations of the flanges and webs of these folds in the same shell sheeting. Such a non-linear relationship between the relative width increments the b_{wr} values of various experimental shell folds at their transverse ends, and the measure of the fold's unit twist angle α_j is shown in Figure 25. The obtained dependencies are functional dependencies, and can be used to calculate the width of the transformed folds depending on the unit twist angles of these folds. The similar nature of the curves indicates the relatively insignificant interdependence between the change in the width of the effectively transformed folds and their height.

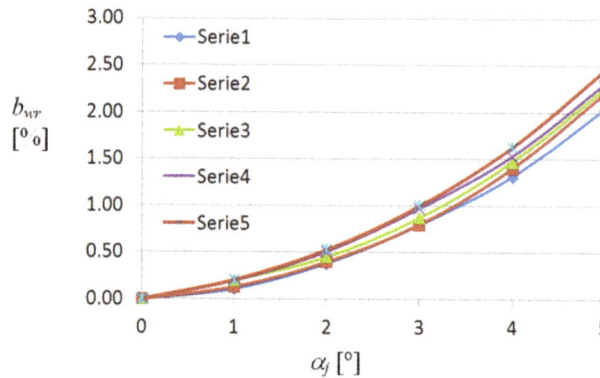

Figure 25. Non-linear relationships between the b_{wr} relative width increments of the fold's crosswise ends and α_j unit angle obtained for various heights of the experimental shell folds: Serie1—50 mm; Serie2—55 mm; Serie3—85 mm; Serie4—136 mm; Serie5—160 mm.

In this step, no new parameter, that is no new independent variable, needs to be adopted. The basic activities of this step are:

- determination of the supporting conditions for all the subsequent folds in the shell sheeting, including the fold's twisting angles, based on the mutual position and shape of the roof directrices,
- calculation of the lengths of supporting lines e_i and f_i for each fold on the basis of these supporting conditions,
- calculation of the arrangement of the supporting points of the fold's transverse ends along each directrix and the total length of directrices e and f,
- determination of the finite number of rulings corresponding to the longitudinal edges of all the shell folds, and
- a possible correction to the shapes of these directrices to obtain the complete coverage of both roof directrices.

The new dependent variables used in this step of the algorithm include:

- the accuracy of the calculations related to the location of the shell fold contraction at its length, and
- the accuracy of the calculations related to the surface area of the smooth shell model of each investigated fold.

7. Parametric Elevation Elements

In the third step of the algorithm, a parametric description of the spatial forms of the considered building roofs and elevations is used. This description also includes the materials from which the roof and elevations are made. Two sufficiently accurate models are shown in Figure 26. The first model, Figure 26a, takes account of the thickness of the roof and elevations, as well as upper, lower, and lateral surfaces of the roof. The other one presents a regular pattern on its elevation walls. The main object of this step is a model taking account of the above-mentioned properties of roof and elevations.

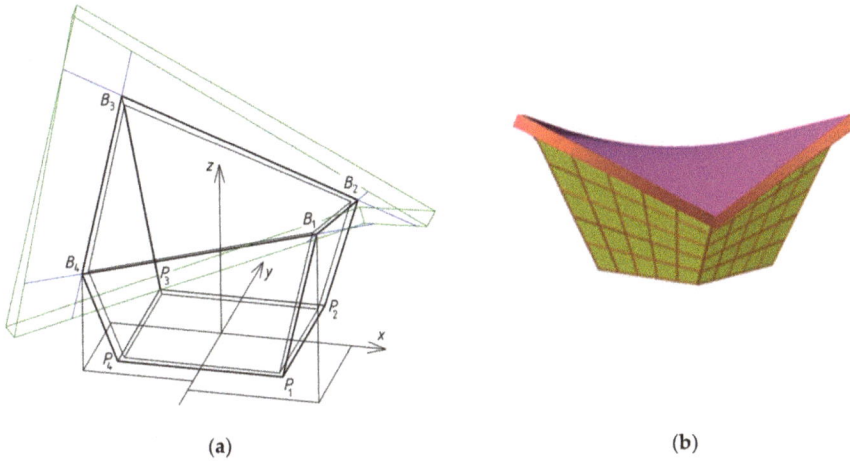

(a)   (b)

Figure 26. Two models presenting: (**a**) the thickness of roof and elevations; and (**b**) the regular pattern on elevations.

The main geometrical elements such as surfaces, planes, lines, and points are the auxiliary objects of this step. The intersections, displacements, and rotations of these elements allow all of the building elements as finite sectors, sides, and edges to be modeled and arranged relative to the building construction axes.

The mutual location of the roof and elevations result from the structure and overall dimensions of the general building form. Therefore, the new parameters used in this step are only:

- ptr_{14}—the thickness of the roof,
- ptr_{15}—the roof overhang outside the outline of elevation walls, and
- ptr_{16}, ptr_{17}—the pitch, position, and inclination of the regular elevation pattern.

8. Example of Shaping Architectural Free Forms

This section presents the geometrical properties of the transformed roof shell roofing for the architectural form that is sought after in this work, the general form of which has been defined using the parameters given in Table 1 in Section 5, and is illustrated in Figure 13.

The shell roof is characterized by non-zero thickness, which is expressed by means of parameter $ptr_{14} = 720$ mm, and the overhang of the eaves outside the outline of the façade, which is expressed by the parameter $ptr_{15} = 500$ mm (Figure 27). The coordinates of its characteristic points and the general form of the entire discussed architectural form are given in Table 3. Points D_{e1}, D_{e2}, D_{f1}, and D_{f2} are additionally selected on the e_g and f_g directrices of the upper roof shell surface Ω_g, and their coordinates are entered with the coordinates of points D_{g1}, D_{g2}, D_{g3}, and D_{g4} as input defining the directrices used in the application of the Rhino/Grasshopper program, which was discussed in Section 6. These values were adopted as the suggested values of the sliders that are presented in Figure 19.

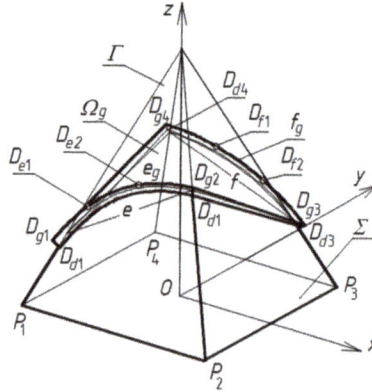

Figure 27. Selected points and lines characterizing the general form of the discussed free form.

Table 3. Parameters achieved for the examined architectural free form from Figure 27.

Vertex	X-Coordinate	Y-Coordinate	Z-Coordinate
P_1	−10,000.0	−10,000.0	0.0
P_2	10,000.0	−10000.0	0.0
P_3	10,000.0	10000.0	0.0
P_4	−10,000.0	10000.0	0.0
B_1	−7190.2	7190.2	6556.2
B_2	4537.3	−4537.3	12,746.2
B_3	7190.2	−7190.2	6556.2
B_4	−4537.3	4537.3	12,746.2
$S_1 = S_2 = H_1 = H_2 = H_3 = H_4$	0.0	0.0	23,333.3
D_{g1}	−8011.4	−7977.4	5988.8
D_{g2}	5009.9	−4885.4	13,203.5
D_{g3}	8011.4	7977.4	5988.8
D_{g4}	−5009.9	4885.4	13,203.5
D_{d1}	−7128.7	−8162.4	5556.9
D_{d2}	5140.9	−5166.9	12,546.4
D_{d3}	7128.7	8162.4	5556.9
D_{d4}	−5140.9	5166.9	12,546.4
D_{e1}	−5349.2	−6561.3	9293.0
D_{e2}	−413.5	−5291.2	12,256.5
D_{f1}	413.5	5291.2	12,256.5
D_{f2}	5349.2	6561.3	9293.0

[1] Values in millimeters.

The discussed roof shell is limited from the top and bottom by two oblique surfaces, the upper one of which is the sought-after model of the transformed folded shell sheeting. This model (Figure 28) was determined using the innovative application built by one of the authors in the Rhino/Grasshopper program.

The calculated values of the unit twist angle α_j defining the supporting conditions, and geometrical properties of the subsequent folds in the discussed shell roof, are tabulated in Table 4.

Figure 28. Smooth model of the upper surface of the shell roof being sought.

Table 4. Parameters describing the subsequent shell folds in the transformed shell whose simplified smooth model is shown in Figure 28.

Shell Fold [no.]	Length of Supporting Line e_i [mm]	Length of Supporting Line f_i [mm]	Fold's Length [mm]	Fold's Unit Twist Angle α_j [°]
1	481	314.2	14,895	3.9834
2	469.5	314.3	14,595	4.037
3	458.6	314.5	14,313	4.0927
4	448,6	314.9	14,047	4.15
5	439	315.4	13,797	4.2081
6	430.1	315.9	13,563	4.2664
7	421.8	316.7	13,344	4.3243
8	414	317.4	13,140	4.3811
9	406.4	318.3	12,950	4.4363
10	399.4	319.3	12,775	4.4893
11	393	320.4	12,615	4.5399
12	386.6	321.6	12,468	4.589
13	381.1	322.8	12,335	4.637
14	375.6	324.4	12,216	4.6835
15	370.7	326.1	12,110	4.7279
16	366.2	327.9	12,018	4.7696
17	361.9	329.8	11,938	4.8082
18	357.7	332.2	11,872	4.8433
19	353.8	334.4	11,819	4.8731
20	350.2	336.9	11,779	4.8963
21	347.1	339.5	11,752	4.9125
22	343.9	342.3	11,738	4.9216
23	340.9	345.5	11,737	4.9234
24	337.9	348.7	11,750	4.9181

Table 4. Cont.

Shell Fold [no.]	Length of Supporting Line e_i [mm]	Length of Supporting Line f_i [mm]	Fold's Length [mm]	Fold's Unit Twist Angle α_j [°]
25	335.4	352.2	11,775	4.9057
26	333	355.5	11,813	4.8862
27	330.6	359.5	11,865	4.86
28	328.6	363.8	11,930	4.8279
29	326.7	368.1	12,008	4.7918
30	324.8	372.7	12,099	4.7525
31	323.2	377.8	12,204	4.7104
32	321.6	383.3	12,322	4.666
33	320.3	389.2	12,455	4.62
34	319.1	395.6	12,601	4.5728
35	318	402.4	12,761	4.5245
36	317	409.6	12,936	4.4737
37	316.1	417.2	13,125	4.4206
38	315.2	425.6	13,330	4.3658
39	314.5	434.3	13,550	4.31
40	314	443.5	13,786	4.2534
41	330.8	427.6	14,026	4.1907
42	312.7	462.3	14,282	4.1288
43	312.5	473.6	14,566	4.0746
44	294.7	513.9	14,880	4.0292

The final architectural free form that is sought is presented in Figure 29. The parameter ptr_{16}, which is defining the position of the subsequent horizontal lines in the regular elevation pattern, is 3000 mm. The parameter ptr_{17}, which is defining the position of the vertical lines of the regular elevation pattern, is four, because of four vertical elevation glass strips.

Figure 29. Two free-form buildings roofed with transformed shells located in a built environment.

The free-form buildings under consideration are visually attractive owing to the suitable shell shapes of roofs coherent with the oblique elevations. It is also noticeable how the color and regularity of elevation pattern affect the attractiveness and harmonious incorporation of the parametric free-form buildings into the built environment. Abramczyk presented a way of adopting fine proportions between the parameters, leading to attractive building free forms [14]. The proposed method requires the designer to possess a certain predisposition to logical and spatial reasoning in the field of shaping free forms, as well as their texture, color, light, and shadow. Obtaining the required effect is conditioned by the individual artistic predispositions of the designer. The method assists the designers in managing their ambitious artistic goals.

9. Conclusions

The novel parametric description of the unconventional building free forms roofed with nominally plane-folded steel sheets transformed into shell sheeting is presented. The algorithm of the innovative method proposed for shaping the aforementioned forms with transformed corrugated shell roofs is based on the description. The method must be supported by the authors' application, which was created in the Rhino/Grasshopper program useful in parametric design

The proposed parametric description and the algorithm based on the description are related to the multidimensional aspects of the architectural free-form design. The use of the above parametric description is presented in detail regarding the example of shaping one relatively simple architectural free form with a transformed shell roof. The visualization of this form, which is shown in the last figure, is the effect of using the above description as well as the computer-aided method based on this description.

The article presents three basic types of the discussed architectural forms. Their general forms have diversified shapes, where their widths change at the height from the building base to the eaves in various ways. One of the presented forms expands along the height of the oblique façade walls from the base to the eaves (Figure 12), while the other contracts (Figure 13). In the third type of the presented forms (Figure 10), the width of the whole shaped building measured between the two opposite façade walls increases, while the width between the other pair of the opposite façade walls decreases in the vertical direction from the base to the eaves. The possibility of determining the various types of architectural forms, in which the elevation edges and planes are inclined to the vertical to various degrees, and roof surface rulings and eaves edges are inclined to the horizontal base plane, is clearly demonstrated. This proves the sensitivity of the proposed method to the harmonization of these forms with the built environments.

The intended effect consisting of creating an attractive unconventional architectural form should be achieved not so much by adopting the values of the proposed parameters, as much as by adopting the proportion between these values and one basic parameter, which is called the reference one. In the case of the chosen architectural form, the reference parameter is ptr_{13} = 20,000 mm, which describes the width of the architectural form. The adopted ratios are $ptr_1/ptr_{13} = ptr_2/ptr_{13} = 1.17$, ptr_i/ptr_{13} for i = three to six, $ptr_j/ptr_{13} = 0.64$ for j = seven to 10, $ptr_{11}/ptr_{13} = 0.090$, $ptr_{12}/ptr_{13} = 0.036$, $ptr_{14}/ptr_{13} = 0.036$, $ptr_{15}/ptr_{13} = 0.025$, and $ptr_{16}/ptr_{13} = 0.15$. The adopted proportions and values make it possible to define the roof and elevation lines, including roof directrices. On the basis of the shape and mutual position of the directrices, the supporting conditions are calculated, and followed by the smooth shell models of the subsequent folds of the roof shell.

The authors' computer application supports these calculations. The application contains two basic conditions determining whether the created simplified smooth models guarantee the effectiveness of the fold's transformations during the assembly of these folds into the calculated places arranged along roof directrices. The first condition concerns the equality of the surface areas of a smooth model of a fold before and after its transformation. The experimental tests and computer analyses have shown that for folds of different profiles, and therefore of different lateral stiffness, the above areas differ relatively little compared with the accuracy of shell modeling. Further detailed experimental tests in this field are necessary in order to develop a function correcting the surface area of each fold after transformation, depending on its lateral stiffness. The second condition concerns the location of the fold's contraction along its length, and has to be rigorously observed, because even a relatively small change in the position of this contraction in relation to the length of the fold results in a significant change in the proportion between the lengths of both its supporting lines *e* and *f*, and this significantly affects the transverse stresses, which is decisive for the value of the fold's initial effort.

As the thickness of each shell roof should be conspicuous, that is, significant in relation to the height of the building, both its upper and lower surfaces can be made up of the transformed corrugated sheeting. Both should usually be determined by means of the method, despite really small differences in the curvature of these surfaces. This results from even small differences in the supporting conditions of the folds of both shell sheetings inducing additional changes in the supporting line length. After the lengths of these changed lines have been added up, major differences arise in both the length of the supporting line of the entire roof shell, as well as in the spacing of the fixing points of these folds along the roof directrices. Additionally, it should be taken into account that these folds are usually supported by additional intermediate directrices at their length.

The architectural form of any designed building should be internally consistent; this means that the shape, position, and orientation of its characteristic straight and curved edges, as well as the flat and curved surfaces of the roof and façade, must be integrated. This integration must be taken into account both at the step of general form specification and when shaping the façade pattern. For this purpose, the proposed description includes parameters that define the regular form of the elevation pattern. In the case of the architectural form selected for discussion, a simple, equal division of each façade wall into horizontal and vertical glass strips separated by lines obtained by dividing each of its four edges into sections of equal length has been used. As the authors' analysis of the division of glass elevation planes into uneven strips or pattern diagonal orientation—which affects the integrity of the architectural form and its sensitivity to the built environment—is not complete yet, its results are not presented in the article.

The authors intend to continue and extend their research to the following areas: (1) the parametric description of free forms of complete buildings and their structures roofed with transformed corrugated shells; (2) the search for rational structural systems dedicated to the buildings under consideration here; and (3) the development of numerical models calibrated on the basis of their experimental research and exhibiting the geometrical and mechanical properties of elastically transformed thin-walled folded shells.

In the first case, the authors propose to analyze the possibilities of joining a few individual free forms of all three previously described types into a single structure with folded or segmented elevations and roof, which can be even more sensitive to the built environments than some complete forms. In the second case, due to the oblique orientation of the edges and surfaces of the façade and roof, it is necessary to adjust the shape of the structural system not only to the shape of the architectural form and its elements, but also to the character and direction of the characteristic load. Building construction has to guarantee an appropriate stiffness of architectural form, especially along the oblique edges of the roof and façade. The authors intend to conduct the analyses of various single-branch and multi-branch forms of structural elements such as poles and roof girders, depending on the architectural form dimensions and the roof span.

Author Contributions: J.A. carried out research and analyses, visualized and interpreted the results, created models and method as well as wrote all sections of the paper. J.A. was the supervisor and the project administrator. A.P. participated in the concept and investigations as well as funding acquisition.

Funding: The resources of the Rzeszow University of Technology.

Conflicts of Interest: The authors declare no conflict of interest.

References

1. Reichhart, A. *Geometrical and Structural Shaping Building Shells Made up of Transformed Flat Folded Sheets*; Rzeszow University of Technology: Rzeszów, Poland, 2002. (In Polish)
2. Foraboschi, P. The central role played by structural design in enabling the construction of buildings that advanced and revolutionized architecture. *Constr. Build. Mater.* **2016**, *114*, 956–976. [CrossRef]

3. Foraboschi, P. Structural layout that takes full advantage of the capabilities and opportunities afforded by two-way RC floors, coupled with theselection of the best technique, to avoid serviceability failures. *Eng. Fail. Anal.* **2016**, *70*, 377–418. [CrossRef]

4. Isler, H. Generating shell shapes by phisical experiments. *Bull. IASS* **2016**, *111*, 53–63.

5. Foraboschi, P. Modeling of collapse mechanisms of thin reinforced concrete shells. *J. Struct. Eng. ASCE* **1995**, *121*, 15–27. [CrossRef]

6. Foraboschi, P. Optimal design of glass plates loaded transversally. *Mater. Des.* **2014**, *62*, 443–458. [CrossRef]

7. Foraboschi, P. Versatility of steel in correcting construction deficiencies and in seismic retrofitting of RC buildings. *J. Build. Eng.* **2016**, *8*, 107–122. [CrossRef]

8. Abramczyk, J. *Shell Free Forms of Buildings Roofed with Transformed Corrugated Sheeting*; Rzeszow University of Technology: Rzeszów, Poland, 2017.

9. Abramczyk, J. An Influence of Shapes of Flat Folded Sheets and Their Directrices on The Forms of The Building Covers Made up of These Sheets. Ph.D. Thesis, Rzeszow University of Technology, Rzeszów, Poland, 11 October 2011. (In Polish)

10. Abramczyk, J. Shape transformations of folded sheets providing shell free forms for roofing. In Proceedings of the 11th Conference on Shell Structures Theory and Applications, Gdańsk, Poland, 11–13 October 2017; Pietraszkiewicz, W., Witkowski, W., Eds.; CRC Press Taylor and Francis Group: Boca Raton, FL, USA, 2017; pp. 409–412.

11. Bathe, K.J. *Finite Element Procedures*; NJ Prentice Hall: Englewood Cliffs, NJ, USA, 1996.

12. Parker, J.E. Behavior of Light Gauge Steel Hyperbolic Paraboloid Shells. Ph.D. Thesis, Cornell University, Ithaca, NY, USA, 1969.

13. Gergely, P.; Banavalkar, P.V.; Parker, J.E. The analysis and behavior of thin-steel hyperbolic paraboloid shells. In *A Research Project Sponsored by the America Iron and Steel Institute*; Report 338; Cornell University: Ithaca, NY, USA, 1971.

14. Abramczyk, J. Parametric shaping of consistent architectural forms for buildings roofed with corrugated shell sheeting. *J. Archit. Civ. Eng. Environ.* **2017**, *10*, 5–18.

15. Obrębski, J.B. Observations on Rational Designing of Space Structures. In Proceedings of the Symposium Montpellier Shell and Spatial Structures for Models to Realization IASS, Montpellier, France, 20–24 September 2004; pp. 24–25.

16. Rębielak, J. Review of Some Structural Systems Developed Recently by help of Application of Numerical Models. In Proceedings of the XVIII International Conference on Lightweight Structures in Civil Engineering, Łódź, Poland, 7 December 2012; pp. 59–64.

17. Grey, A. *Modern Differential Geometry of Curves and Surfaces with Mathematica*, 4th ed.; Champman & Hall: New York, NY, USA, 2006.

18. Abramczyk, J. Method for Parametric Shaping Architectural Free Forms Roofed with Transformed Shell Sheeting. *JCEEA* **2014**, *61*, 5–21. [CrossRef]

19. McDermott, J.F. Single layer corrugated steel sheet hypars. *Proc. ASCE J. Struct. Div.* **1968**, *94*, 1279–1294.

20. Egger, H.; Fischer, M.; Resinger, F. *Hyperschale aus Profilblechen*; Der Stahlbau, H., Ed.; Ernst&Son: Berlin/Brandenburg, Germany, 1971; Volume 12, pp. 353–361.

21. Davis, J.M.; Bryan, E.R. *Manual of Stressed Skin Diaphragm Design*; Wiley: Granada, UK, 1982.

22. Petcu, V.; Gioncu, D. Corrugated hypar structures. In Proceedings of the I International Conference on Lightweight Structures in Civil Engineering, Warsaw, Poland, 25–29 December 1995; pp. 637–644.

23. Reichhart, A. Principles of designing shells of profiled steel sheets. In Proceedings of the X International Conference on Lightweight Structures in Civil Engineering, Rzeszow, Poland, 5–6 December 2004; pp. 138–145.

24. Reichhart, A. Corrugated Deformed Steel Sheets as Material for Shells. In Proceedings of the I International Conference on Lightweight Structures in Civil Engineering, Warsaw, Poland, 25–29 December 1995.

25. Kiełbasa, Z. Computational shaping of the twisted folded sheet. In Proceedings of the Conference on New Challenges in Constructions Structural Shaping Thin-walled Corrugated Constructions, Rzeszów, Poland, 17–21 May 2000. (In Polish)

26. Prokopska, A. Creativity Method applied in Architectural Spatial cubic Form Case of the Ronchamp Chapel of Le Corbusier. *J. Transdiscip. Syst. Sci.* **2007**, *12*, 49–57.

27. Abramczyk, J. Building Structures Roofed with Multi-Segment Corrugated Hyperbolic Paraboloid Steel Shells. *J. Int. Assoc Shell Spat. Struct.* **2016**, *2*, 121–132. [CrossRef]

28. Wilk, A. Project of a Car Showroom Roofed with Transformed Folded Steel Sheeting. Bachelor's Thesis Diploma Thesis, Rzeszow University of Technology, Rzeszów, Poland, 2016. (In Polish)

29. Wojtuń, A. Design of Transformed Folded Shell Roof Sheeting over Cyclist Service Area. Bachelor's Thesis Diploma thesis, Rzeszow University of Technology, Rzeszów, Poland, 2016. (In Polish)

30. Prokopska, A.; Abramczyk, J. Innovative systems of corrugated shells rationalizing the design and erection processes for free building forms. *J. Archit. Civ. Eng. Environ.* **2017**, *10*, 29–40. [CrossRef]

31. Abramczyk, J. Principles of geometrical shaping effective shell structures forms. *JCEEA* **2014**, 5–21. [CrossRef]

buildings

MDPI

Article

The Structural Effectivity of Bent Piles in Ammatoan Vernacular Houses

Wasilah Wasilah

Architectural Engineering, Faculty of Science and Technology, Universitas Islam Negeri Alauddin, Makassar 92113, Indonesia; wasilah@uin-alauddin.ac.id; Tel.: +62-812-410-0969

Received: 30 December 2018; Accepted: 2 February 2019; Published: 10 February 2019

Abstract: Ammatoa Kajang vernacular houses are buildings that have existed for a hundred years as residential house buildings. These traditional houses are unique in their use of bent piles. This research examines the strength of the structural system of Ammatoan vernacular houses based on said houses' ability to adapt to various environmental conditions and natural phenomena. This study seeks to enrich these studies by examining the specific structural strength of these buildings. In the face of modernization and extreme climate change, the continued existence of such traditional houses has been threatened. Disaster may strike at any time, and as such we must explore the structural strength of their structures to predict these buildings' ability to endure such events. This research applies an interpretative model to explore the structural system, using a load test to examine the houses' structural strength. Although such a model assumes that each building has the same pitch, each house has its own pitch. Therefore, the measurement results cannot be applied generally to describe the structural strength of every Ammatoan house. This research also notes that the pin joint system, material selection, and application of a grounded foundation are factors that promote these buildings' continued endurance and ability to withstand earthquakes.

Keywords: bent piles; structural systems; vernacular houses; Ammatoan houses; building structure; effectivity; strength

---

1. Introduction

The structure of a building refers to the constitution of its elements, which bears the load of the main construction; the examination of a structure focuses on elements that actually bear the load, without considering whether they appear to do so. Generally, such a structure consists of a foundation, walls, columns, floors, and trusses [1,2].

Traditional architectures tend to rely on simple structures, with their stability determined by traditional peoples' empirical experience, intuitive knowledge, and attempts to pass knowledge from generation to generation [3–6]. Structure is closely linked to the anatomy of a building, i.e., the sub-structure and upper structure. Construction, meanwhile, is closely linked to the methods, techniques, and means through which elements are bound, lifted, connected, etc [7].

The Ammatoans, also known as the Kajang, have maintained a traditional lifestyle for hundreds of years and synergized with their local natural environment (see Figure 1). This consistency in traditional values significantly influences the structures and forms of Ammatoan architecture [8,9]. As with other traditional buildings in South Sulawesi, Ammatoan vernacular houses are stilt houses. They are characterized, however, by a relative homogeneity, simplicity, small room dimensions, use of natural materials, and lack of ornamentation or other signs of social stratification [10].

Figure 1. Ammatoan Wood House.

Another key characteristic of traditional Ammatoan houses is the bent pile structural system. This system has been used by the Ammatoans for hundreds of years, and each pile is capable of standing for decades (or even centuries) without any instability. This becomes interesting in relation to further investigations about the strength of building structures and how they can last long.

These buildings are built following the principle of *mappasituppu* [11], which invokes the natural characteristics of the trees used, requires the use of bent piles, and sets a specific location for each pile. This structural system has remained the primary one in Kajang, and can support Ammatoan vernacular houses—as well as the various activities within—for over a hundred years. It is this system that is examined in this research.

All parts of Ammatoan houses are made entirely from wood. The columns, beams, walls, stairs, and all the floors use wood material, while the roof cover is made of zinc and thatch material. As such, these timber structures have relatively little weight, providing these residential buildings with a clear advantage. Consequently, these buildings have a minimal shear force.

Wood is a natural material, which has three main axes [12]. Along the primary axis, the strength and stiffness are superior; in terms of the strength–material density ratio, it is even stronger than other materials. However, wood is relatively weak and soft along the two others axes, and as such it can crack and cause structural failure. Since wood has different mechanical properties along its three axes, a nonlinear analysis is required.

Through their physical structure, construction, and material, traditional or vernacular houses convey different values [13] (Rapoport, 1969) (see Figure 2). As such, previous studies of Ammatoan houses have focused primarily on the integration of Ammatoan cultural values in their architecture. This study seeks to enrich these studies by examining the specific structural strength of these buildings. In the face of modernization and extreme climate change, the continued existence of such traditional houses has been threatened [14]. Disaster may strike at any time, and as such we must explore the structural strength of their structures to predict these buildings' ability to endure such events [15,16]. The key contribution is the recognition of the specific structural strength of these vernacular Ammatoa houses within modernization and extreme climate change as a basis for consideration in building similar house buildings in South Sulawesi.

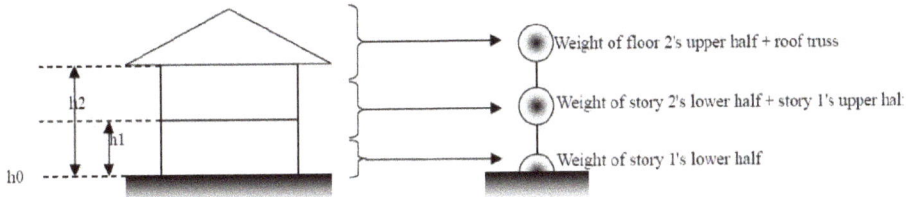

Figure 2. Weight modeling of each floor of a wood house model. (Suwantara and Suryantini, 2014).

2. Research Methods

This study examines the structural system of Ammatoan vernacular houses, as influenced by the culture and beliefs of the local society. This research applies an interpretative model to explain the meaning of the bent piles and their output in Ammatoan vernacular houses. An analysis was done using SAP2000 software to observe the strength and endurance of these bent piles. In the modeling and analysis, the finite element method was used.

This research involves the following:

(a) The material and joints used in Ammatoan vernacular houses were identified, in order to recognize the material types before a mechanical testing.
(b) A mechanical testing was conducted, applying Wood. It was found that the mechanical strength is influenced by flexural strength, compressive strength, tensile strength, shear, and modulus of elasticity (MOE).
(c) The results of the mechanical testing were examined using a numerical analysis, as well as a wind load and live load. This was used to predict the strength of the bent pile structural system, including its ability to endure earthquakes.

Ammatoan vernacular houses are made using local wood material, primarily bitti wood (Vitex cofassus). However, information about this wood is limited, and the wood available is of limited diameter as it comes from young trees. Therefore, it is necessary to determine the characteristics and properties of bitti wood. Bitti wood is classified as Class II in terms of durability [17] (Forest Botanical Departments, 1972), and has an elastic modulus of 12,000 kg/m². The volume of the load structure application complies with the terms of the *Procedure for Planning the Wood Structures of Buildings* (1987). Its wind load is 40 kg/m²; live load is 200kg/m²; and wood volume is 1000 kg/m³ [18–20]. This information was entered into the SAP2000 model. The Structure Modeling, Data Input (see Table 1) includes:

1. Material Properties: the quality of the material used is Wood. Beams and Columns: E13 (fc '= 28 MPa)
2. Structure Dimension: the structural dimensions that will be used in this Ammatoa house structure model are:

    a. Beams B1 = 300 × 300 mm
    b. Beam B2 = 200 × 300 mm
    c. Column C1 = 500 × 500 mm

3. Loading: the loading structure used in this model is:

    a. Dead Load (DL)
        - Reinforced concrete: 1000 kg / m³
    b. Life Expenses (LL)
        - Roof Floors: 50 kg / m²

- Other floors: 150 kg / m^2

c. Earthquake Load

The earthquake load used in this study is the earthquake load time history Bulukumba, South Sulawesi taken from the U.S. Geological Survey (USGS) Earthquake Hazards Program.

Table 1. Data Input for Structure Modeling in SAP 2000.

No.	Item		Value	Unit
1	Material Properties		28	Mpa
2	Structural dimensions	Beams B1	300 × 300	Mm
3		Beams B1	200 × 300	Mm
4		Column C1	500 × 500	Mm
5	Dead Load (DL)	Reinforced concrete	1000	kg/m^3
6	Life Expenses (LL)	Roof Floors	50	kg/m^2
7		Other floors	150	kg/m^2
8	Wind load		40	kg/m^2

Based on research by Hapid (2010), bitti wood has a consistency of 11.17–17.88% vessels, 57.49–70.32% fibers, 10.74–19.30% rays, and 5.40–8.98% parenchyma. The cell measurement found an average fiber length of 0.86–1.44 mm, average fiber diameter of 17.81–20.59 μm, average lumen diameter of 11.93–13.86 μm, and average cell wall thickness of 2.61–4.02 μm. It had a green and air-dry moisture content of 59.32–110.22% and 9.66–20.82%, respectively. The specific gravity in green, air-dry, and oven-dry conditions were 0.48–0.70, 0.48–0.75, and 0.50–0.80, respectively. The radial, tangential, and longitudinal shrinkage of the wood from a green to an oven-dry condition was 3.64–7.44%, 1.79–4.32%, and 0.18–0.49%, respectively, with a T/R ratio of 1.42–2.03. The radial, tangential, and longitudinal swelling of the wood from an oven-dry to a wet condition was 3.673, 4.374%, and 0.342%, respectively, with a T/R ratio of 1.32–3.26. The static bending strength, MOE, and MOR were 460.37–803.17 kg/cm^2, 67.75–97.22 (x 103 kg/cm^2), 634.13–1046.39 kg/cm^2, respectively [21].

3. Results

The upper structure refers to certain elements of the construction: columns, load-bearing beams, and roof frames. In principle, the roof of a building serves the same function as its walls, namely to cover the building and protect it from the elements (heat, cold, etc.). The light construction refers to the construction system used in tropical areas.

The structure used for Ammatoan traditional houses is made primarily with bitti wood, a locally available material. It relies solely on a peg jointing system, including for its roof construction. The main piles of the structure are embedded about tone meter into the ground. They are arranged in a grid, measuring 6 × 9 meters. Two types of piles are used in the construction: the piles that support the roof trusses (average length: ± 4 m) and the piles that support the floor beams (average length: ± 1.5 m). All of the piles are made with bitti wood, with a diameter of 15 to 20 cm. These wooden piles are made using entire tree trunks, and are not straight but bent or curved.

The floors of Ammatoan houses are supported by beams that measure 10 to 15 cm in diameter, with wooden boards providing the floor itself. The second-story beams are close together (± 40 cm) and connected to the primary pile. These beams are covered with boards, which are nailed to the beams.

The strength of a bent pole in the Ammatoan house structure was tested using SAP2000, and it was analyzed through FE models. The output analysis shows the following advanced solver:

(a)   Number of joints = 105

(b)   Number of frame/cable/tendon elements = 188

(c)   Number of shell elements = 24

(d)   Number of load patterns = 3

(e)   Number of the sacceleration load = 6

(f)   Number of load cases = 5

Stiffness At Zero (Unstressed) Initial Conditions

(a)   Number of stiffness degrees of freedom = 510

(b)   Number of mass degrees of freedom = 255

(c)   Maximum number of eigen modes sought = 12

(d)   Minimum number of eigen modes sought = 1

(e)   Number of residual-mass modes sought = 0

(f)   Number of subspace vectors used = 24

(g)   Relative convergence tolerance = 0.000000001

(h)   Found mode 1 of 12: EV = 1794.4866, f = 6.742023, T = 0.148323

(i)   Found mode 2 of 12: EV = 2993.5333, f = 8.707875, T = 0.114839

(j)   Found mode 3 of 12: EV = 4311.4325, f = 10.450352, T = 0.095691

(k)   Found mode 4 of 12: EV = 4364.2055, f = 10.514115, T = 0.095110

(l)   Found mode 5 of 12: EV = 5455.0936, f = 11.754962, T = 0.085070

(m)   Found mode 6 of 12: EV = 6835.7308, f = 13.158689, T = 0.075995

(n)   Found mode 7 of 12: EV = 7071.0202, f = 13.383237, T = 0.074720

(o)   Found mode 8 of 12: EV = 7383.9175, f = 13.676140, T = 0.073120

(p)   Found mode 9 of 12: EV = 7441.7295, f = 13.729574, T = 0.072835

(q)   Found mode 10 of 12: EV = 7536.3680, f = 13.816600, T = 0.072377

(r)   Found mode 11 of 12: EV = 25618.453, f = 25.473967, T = 0.039256

(s)   Found mode 12 of 12: EV = 32740.592, f = 28.798071, T = 0.034725

This structural analysis found that the piles of the Ammatoan vernacular houses are extremely strong, more than capable of handling the live load and dead load, with a roof structure that can readily resist the wind load. The third level of the structure, or the house ceiling, has beams that support the carrying of the load, as found in the structural test. Therefore, the ceiling may require frequent replacements to ensure the maximal performance of the house structure.

According to the load analysis, the results of which are shown in Figure 3, all piles can properly carry the load. These piles, cyan in the figure above, have the strength to support the building for a long period of time. Field research found that several Ammatoan vernacular houses have been standing for more than 80 years. The piles of these houses have never been replaced, and have been able to consistently carry the load without any indication of damage. However, some supporting beams are colored red in this figure, indicating that these beams tend to lack the ability to handle their load. Many social activities are held in the body of the house, such as weddings, funerals, discussion, etc. Consequently, there is a high probability that these beams require replacement. The Fmax diagram of the load, shown in Figure 4 below, shows the influence of various factors.

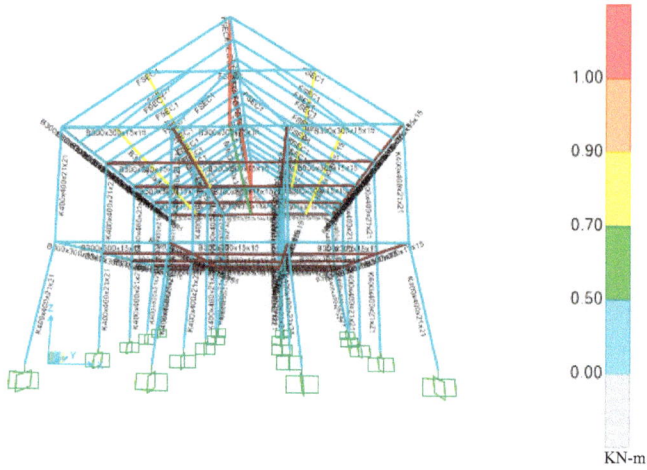

Figure 3. The structural load of an Ammatoan vernacular house.

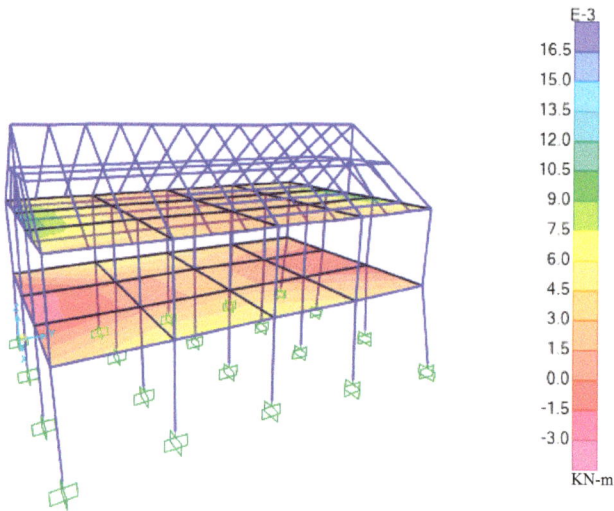

Figure 4. Fmax diagram.

An Fmax diagram depicts the maximum force per length unit in the middle of an element. Its main orientation is presented in such a way that the shear force per length unit is zero. The second floor, or the house body, has a lower main force load than the third level. This can be observed in the color gradation of Figure 4. The relative virtual work (RVW) diagram in Figure 5 shows the virtual work energy as an indication of the force and displacement load pressure.

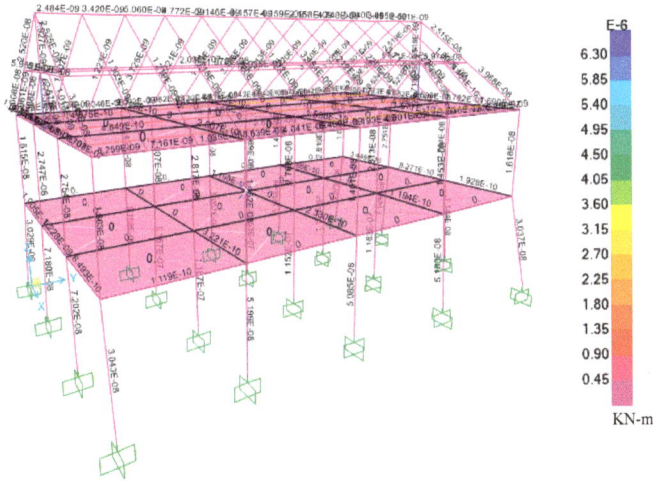

Figure 5. RVW (Relative virtual work) Output Model.

The relative virtual work analysis presents the virtual work percentage of an element relative to the balance of the structural elements. It has the benefit of decreasing the structural deflection by indicating the element with the highest energy percentage, and the deflection is significantly influenced if the stiffness is modified.

Figure 5 depicts the virtual work energy percentage of Ammatoan vernacular houses, with the pressure caused by the pattern of the force and load displacement. This condition explains why, when the percentage of the element stiffness is higher than that of the structural deflection, it is significantly more influential than at a lower percentage.

The analysis found that the element stiffness is fairly low, and that the structural modification is not significantly influenced by the load (dead load, live load, wind load, and earthquake load). These results reflect the development principle of Ammatoan houses, which is oriented toward sustainability. The use of natural forms in piles indicates an elemental stiffness without any modification. As a result, the Relative Virtual Work results indicate that the bent pile structure functions better than a modified pile.

Such vernacular houses have proven to stand firm during earthquakes and other natural disasters. This study explains its results through an analysis of output values, ranging in value from 9.3 to 10 (Figure 6). Referring to an earthquake magnitude estimate for the Bulukumba area (where the Ammatoans live) by the U.S. Geological Survey (USGS) Energy Resources Program, this structure will handle earthquakes without collapsing or experiencing severe damage. However, in earthquakes, such houses will shake and cause an apparent motion of 30–60 cm (Figure 7). The red line in Figure 7 shows the initial position of the Ammatoan living room, while the blue line shows the apparent motion during earthquakes. The difference between the red and blue line on the second story (living area) is about 30–35 cm, while the difference for the third story (attic) is about 40–60 cm.

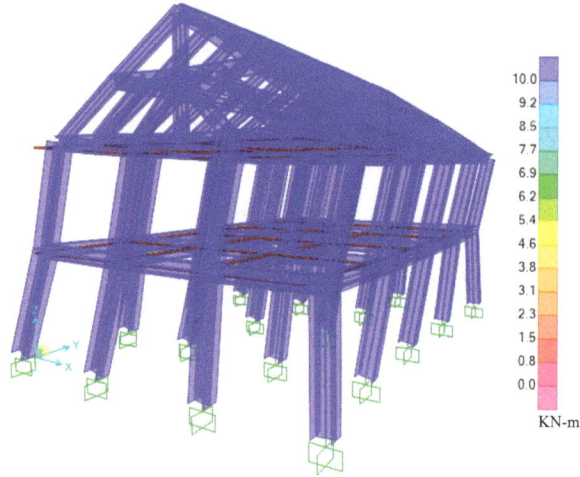

Figure 6. Illustration of the joint system of an Ammatoan vernacular house.

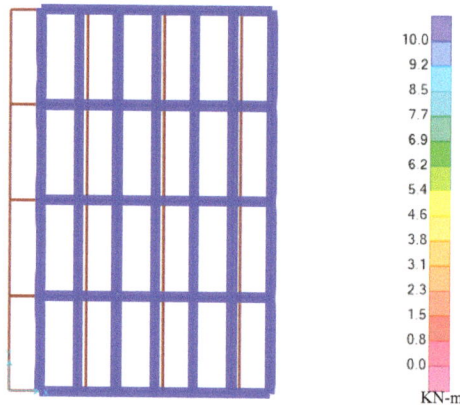

Figure 7. Illustration of the joint system of an Ammatoan vernacular house.

The primary success of Ammatoan vernacular architecture lies in its ability to withstand climate, time, and earthquakes, a situation influenced by the pin joint system, grounded foundation engineering, and pile technique. These three factors are basic elements of Ammatoan residential buildings, as stated in Pasang ri Kajang, the guideline for local life.

4. Discussion

The joints of Ammatoan vernacular houses take into consideration the pedestal foundation, the free positioning, and the stiffness. The pedestal foundation of Ammatoan vernacular houses comes from the wood piles being sunk 1 m into the ground. Ammatoan vernacular houses use both rigid joints and free positioning. Rigid joints are used to connect pillars and beams on the first floor, as well as on the *padongko* beam. Free joints are used in the floor system and to support the ceiling under the roof. In Figure 8 shows the frame of an Ammatoan house.

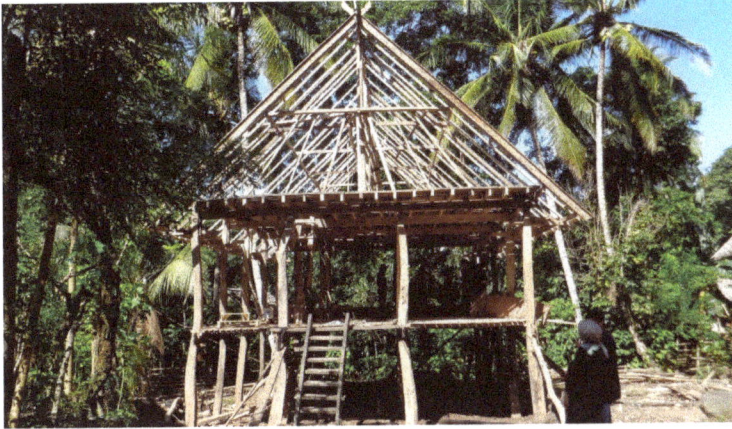

Figure 8. Frame of an Ammatoan house.

The loading on the attic of the house shows that the strength of the roof structure is rather weak, so it is not recommended for accommodating the maximum load. The overall strength of bitty wood has approached its true strength so that its utilization is far more maximal. The obstacle is the beam connection system which does not adjust to the loading pattern of buildings and activities in the building.

House construction in Ammatoan communities may be categorized as environmentally friendly, because houses utilize such natural materials as nipa palm leaves (for roofing), palm fibers and rattan (for binding), bamboo (for floor material), and bitti wood (for walls). Most Ammatoan vernacular houses use timber as their primary material. To build the house, three dowel beams (*padongko*) are arranged horizontally, from the left to the right side of the house. A large horizontal beam is positioned from the left to the right side, and bound to one column (*latta'*) above the house.

The supporting piles are formed from sections or cut trees. Usually, piles in structural systems are square sections, made using raw logs that are carved without any clear object. In Ammatoan vernacular houses, the structural system that supports the main floor consists of bent piles (*benteng*), as well as *unebba, besere, arateng, pattolo*, and flooring. The floorboard size is determined by the size of available material.

A primary component of the roof structure is the truss. Generally, a wood truss construction is a cantilever construction, the primary supporting structure of the roof. A wooden truss construction requires no deformation, especially after implementation. For the roof load to be carried by wooden trusses through the girders, joints must be positioned precisely. Additionally, there must be no bending stress on the bar; only ordinary compressive and tensile stress.

A wooden truss involves wooden beams of a specific dimension that are assembled into isosceles triangles [22]. The trusses are positioned in beam rings, using specific angles. The frame structure may use wood or roller joints (wall plates or respective columns). There are two types of joint systems used in Ammatoan vernacular houses: joints between the piles (*benteng*) and floor beams (*arateng*), and joints between the piles (*benteng*) and roof beams (*padongko*).

There are two types of joints used in Ammatoan houses, while the column system used is a deep pile system (see Figure 9). This simple technology reflects the principle of synergizing with nature (creating balance, sustainability, harmony between man and nature, and ancestral knowledge). In the architecture of the Ammatoan people, these principles have been realized through trial and error. As a result, the Ammatoans have found a form and structure able to carry loads stably for decades.

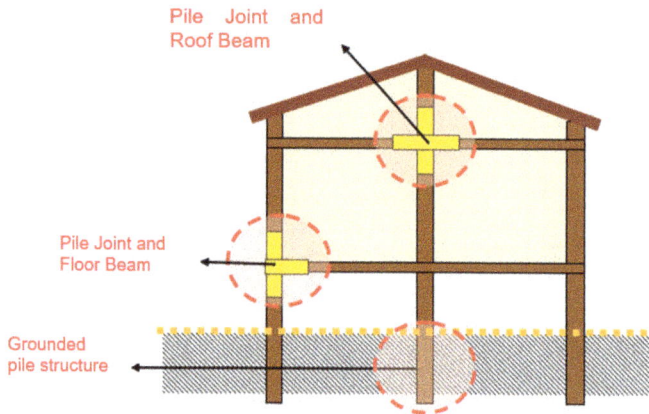

Figure 9. Illustration of the joint system of an Ammatoa vernacular house.

5. Conclusions

(a) A natural understanding of building materials can be combined with traditional expertise to present a structurally sound construction that is also environmentally conscious.

(b) The testing for structural strength applied an assumption of equal pile pitch. However, each Ammatoan house has a different level of pitch. Therefore, the measurement results cannot be generalized to describe the structural strength of every house in Kajang; rather, the results are only applicable to the sample used in this research.

(c) The natural forms and characteristics of tree trunks has led to wood becoming a main building material, a joint system, and a foundation system, and this significantly determined the structural strength of Ammatoan stilt houses. Moreover, the pin joint system, grounded foundation, and selection of piles are primary factors in the success of Ammatoan architecture.

Funding: This research received no external funding.

Acknowledgments: The author would like to express her gratitude to the staff of the Architectural Engineering Laboratory, Faculty of Science and Technology, Alauddin Islamic State University, for sharing their wisdom during the course of this research.

Conflicts of Interest: The authors declare no conflict of interest.

References

1. Frick, H. *Seri Strategi Arsitektur 1 – Pola Struktural dan Teknik Bangunan di Indonesia [Architectural Strategy Series 1 - Structural Patterns and Building Engineering in Indonesia]*; Penerbit Kanisius: Yogyakarta, Indonesia, 1997.

2. Rifai, A.J. *Perkembangan Struktur dan Konstruksi Rumah Tradisional Suku Bajo di Pesisir Pantai Parigi Moutong Developments in the Structure and Construction of Traditional Bajo Houses along the Coast of Parigi Moutong Beach*; Tadulako University: Palu, Indonesia, 2010.

3. Burley, A.L.; Enright, N.J.; Mayfield, M.M. Demographic response and life history of traditional forest resource tree species in a tropical mosaic landscape in Papua New Guinea. *J. For. Ecol. Manag.* **2011**, *262*, 750–758. [CrossRef]

4. Fajrin, J.; Handayani, T.; Anshari, B.; Rofaida, A. *Pengembangan Desain dan Konstruksi Rumah Panggung Sebagai Rumah Murah Berbasis Teknologi Kayu Laminasi Developments in the Design and Construction of Cheap Stilt Houses Using Laminated Wood Technology*; Research Report; Mataram University: Mataram, Indonesia, 2008.

5. Hildayanti, A.; Suriadi, N.A.; Santosa, H.R. Analysis of housing areas with a sustainable community approach. *Int. J. Sci. Eng. Res.* **2014**, *5*, 1511–1517.

6. Luthan, P.L.A. *Pengembangan Konsep Rumah Tinggal Tradisional Mandailing di Sumatera Utara [Conceptual Development of Traditional Mandailing Houses in North Sumatra]*; Gunadarma University: Depok, Indonesia, 2015.

7. Wiryomartono. *KONSTRUKSI KAYU Jilid 1 WOOD CONSTRUCTION Volume 1*; Yayasan Penerbitan FIP-IKIP: Yogyakarta, Indonesia, 1976.

8. Rimang, S.S. *Sejarah Kajang [History of Kajang]*; Lentera Kreasindo: Makassar, Indonesia, 2016.

9. Tika, M. *Ammatoa, Makassar: Lembaga Kajian dan Penulisan Sejarah Budaya Sulawesi Selatan [Ammatoa, Makassar: Institute for the Studies and Writing of the Cultural History of South Sulawesi]*; Pustaka Refleksi: Makassar, Indonesia, 2015.

10. Wasilah, W.; Hildayanti, A. Filosofi penataan ruang spasial vertikal pada rumah tradisional Saoraja Lapinceng Kabupaten Barru [Philosophy of vertical spatial planning in Saoraja Lapinceng traditional houses, Barru Regency]. *J. RUAS (Rev. Urban. Archit. Stud.)* **2017**, *14*, 70–79. [CrossRef]

11. Hartawan; Suhendro, B.; Pradipto, E.; Kusumawanto, A. Perkembangan sistem struktur bangunan rumah Bugis Sulawesi Selatan [The development of the structural system of the Bugis houses of South Sulawesi]. In Proceedings of the 5th Annual Engineering Seminar (AES 2015), Free Trade Engineers: Opportunity or Threat, Faculty of Engineering, Gadjah Mada University, Yogyakarta, Indonesia, 12 February 2015; pp. 51–60.

12. Breyer, D.E. *Design of Wood Structures*; McGraw-Hill: New York, NY, USA, 2007.

13. Rapoport, A. *House Form and Culture*; Prentice-Hall: Englewood Cliffs, NJ, USA, 1969.

14. Setijanti, P.; Silas, J.; Firmaningtyas, S.; Hartatik. *Eksistensi Rumah Tradisional Padang Dalam Menghadapi Perubahan Iklim dan Tantangan Jaman The Existence of Traditional Padang Houses in the Face of Climate Change and the Changing Times*; Sepuluh November Institute of Technology: Surabaya, Indonesia, 2012.

15. Suwantara, I.K.; Suryantini, P.R. Kinerja sistem struktur rumah tradisional Ammu Hawu dalam merespon beban seismik Performance of the Ammu Hawu structural system in responding to seismic loads. *J. Permukim.* **2014**, *9*, 102–114.

16. Wasilah, S.; Fahmyddin, T. The advancement of the built environment research through employment of structural equation modeling (SEM). In *IOP Conference Series: Earth and Environmental Science*; IOP Publishing: Bristol, UK, 2018.

17. Forrest Product Laboratory. *Wood as an Engineering Material (Wood Handbook)*; USDA: Washington, DC, USA, 2010.

18. Mehta, K.C. *Wind Load: Guide to the Wind Load Provisions of ASCE 7-10*; ASCE Press: Reston, WV, USA, 2013.

19. Dogan, M. Seismic analysis of traditional buildings: Bagdadi and Himis. *Anadolu Univ. J. Sci. Technol. A Appl. Sci. Eng.* **2010**, *11*, 35–45.

20. National Standardization Agency of Indonesia (BSN). *Tata Cara Perencanaan Struktur Kayu Untuk Bangunan Gedung Procedure for Planning the Wood Structures of Buildings*; SNI-03-xxxx-2000; National Standardization Agency: Bandung, Indonesia, 2012.

21. Hapid, A. Struktur Anatomi dan Sifat Fisika-Mekanik Kayu Bitti (Vitex Cofassus Reinw) Dari Hutan Rakyat Yang Tumbuh di Kabupaten Bone dan Wajo Sulawesi Selatan The Anatomical Structure and Physical-Mechanical Properties of Bitti Wood (Vitex Cofassus Reinw) from Community Forests Grown in Bone and Wajo Regencies, South Sulawesi. Master's Thesis, Faculty of Forestry, Gadjah Mada University, Yogyakarta, Indonesia, 2010.

22. Ludwig, S. *Kontruksi Kayu Wood Construction*; Erlangga: Jakarta, Indonesia, 2010.

buildings

MDPI

Review

Design and Fabrication of a Responsive Carrier Component Envelope

Teng-Wen Chang [1,*,†,‡], Hsin-Yi Huang [1,‡] and Sambit Datta [2]

[1] College of Design, National Yunlin University of Science and Technology, Taipei 10607, Taiwan; sherry.hhy@gmail.com
[2] School of Electrical Engineering, Computing and Mathematical Sciences, Curtin University, Bentley, WA 6102, Australia; Sambit.Datta@curtin.edu.au
* Correspondence: tengwen@yuntech.edu.tw; Tel.: +886-5-534-2601 (ext. 6510)
† Current address: 123 University Road, Section 3, Douliou, Yunlin 64002, Taiwan, R.O.C.
‡ These authors contributed equally to this work.

Received: 28 February 2019; Accepted: 11 April 2019; Published: 15 April 2019

Abstract: Responsive architecture comprises the creation of buildings or structural elements of buildings that adapt in response to external stimuli or internal conditions. The responsiveness of such structures rests on addressing constraints from multiple domains of expertise. The dynamic integration of geometric, structural, material and electronic subsystems requires innovative design methods and processes. This paper reports on the design and fabrication of a responsive carrier component envelope (RCCE) that responds by changing shape through kinetic motion. The design of the RCCE is based on geometry and structure of carrier surfaces populated with a kinetic structural component that responds to external stimuli. We extend earlier prototypes to design a modular, component-driven bottom-up system assembly exploring full-scale material and electronic subsystems for the expansion and retraction of a symmetric polar array based on the Hobermann sphere. We test the kinetic responsiveness of the RCCE with material constraints and simulate responses by connecting the adaptive components with programmable input and behavior. Finally, a concrete situation from practice is presented where 16 fully-functional components of the adaptive component are assembled and tested as part of an interactive public placemaking installation at the Shenzhen MakerFaire Exhibition. The RCCE experimental prototype provides new results on the design and construction of an adaptive assembly in system design and planning, choice of fabrication and assembly methods and incorporation of dynamic forms. This paper concludes that the design and assembly of an adaptive structural component based on RCCE presents results for designing sensitive, creative, adaptable and sustainable architecture.

Keywords: responsive architecture; kinetic envelope; adaptive design; interactive architecture; moveable facade components; carrier component structures; sensor interaction; digital fabrication

---

1. Introduction

Buildings or elements of buildings that offer adaptive features with the ability to adjust to changes in environmental conditions or external stimuli are termed responsive. Responsive envelopes act as an inter-media [1] between inside and outside, triggering changes that enhance the awareness and experience of place as well as performing functions such as modulation of thermal comfort or lighting. Responsive components can be defined as "all those elements of the building that adapt to the needs of people as well as changes in the environment" [2]. The most common embodiment of responsive envelopes are structures or architectural elements that adapt to changes in climatic variables.

These components may be high tech systems that employ sensor networks and actuators to monitor the environment and automate control of operable building elements. The Tower of Winds [3]

is a cylindrical urban installation, clad in perforated aluminium panels. The responsiveness of the cylindrical envelope is achieved at night by translating the varying sound and wind levels into light through computational methods and translating these sensor data into powering different light emitting devices. The variability of the environment is thus directly made visible in the architecture such that changes in wind speed and noise levels reflect changes in the cylindrical envelope. In this way, the Tower of Winds is constantly transforming, with its small lamps changing colors according the surrounding sounds and its neon rings rippling according to the winds of the city. A second line of development includes dynamic or kinetic structures that respond by reconfiguring their physical shape or form. Aegis hyposurface [4] is a dynamic mechanical wall surface that deforms in response to stimuli. The prototype of faceted metallic panels deform physically as a real time response to electronic stimuli from the environment (movement, sound, and light). Each component of the surface is driven by a bed of pneumatic pistons, generating real-time dynamic terrains.

The design and assembly of responsive structures requires new levels of integration across geometric, structural, material and electronic subsystems. This paper reports on the design and fabrication of a responsive carrier component envelope (RCCE). In this paper, we address the design and fabrication of responsive envelopes that responds to sensing people by changing shape through kinetic motion. The design of the RCCE is based on geometry and structure of carrier surfaces populated with a kinetic structural component that responds to external stimuli. We extend earlier prototypes to design a modular, component-driven bottom up Design-For-Assembly (DFA) system exploring full-scale integration of material, structural, kinetic and electronic subsystems. We present the expansion and retraction of the symmetric polar array based on the Hobermann sphere, test the kinetic responsiveness of the RCCE with material constraints and simulate responses by connecting the adaptive components with programmable sensor input and real-time dynamic behavior. Finally, we prototype and assemble an aggregation of fully functional components as part of an interactive public placemaking installation.

1.1. Background

With the evolution of information technology and smart materials, new categories of responsive intelligent skins have been proposed. Responsive envelopes have been broadly categorized into media facades [5], dynamic envelopes [6,7] and interactive systems [8,9]. The role of the architectural skin as a responsive element [10,11] is a central metaphor in both traditional responsive building facades [12] as well as emergent intelligent facade design [13].

Advances in flexible, configurable geometric representations [14–16] have been central to the design of responsive components. Flexible parametric models permit both design space exploration [17,18] and mapping of digital models to physical infrastructure [19]. Through inexpensive hardware components and embedded sensor systems, virtual simulations can be connected directly to physical models [20]. Sophisticated human–computer interaction paradigms such as mixed-initiative [21] dynamic interfaces [22] and distributed interaction [23] enable responsive components and their interaction with complex environmental changes as well as internal and external stimuli. Furthermore, three-dimensional printing, laser cutting, and desktop milling allow the digital fabrication of components [24,25] for rapid assembly and prototyping [26].

Sensors are commonly used to track indoor and outdoor environmental variables as well as recognize activity patterns and spatial distributions. Kinetic Architecture present concepts of responsiveness where secondary environmental systems can be attached to a primary structural system combined into a collective behavioral system [27]. The stochastic rotation of tiles [28] and the kinetic behavior of origami techniques with shape memory alloy actuators [29] and dynamic skins [30] have been developed. Structures that respond to lighting and energy optimization have been proposed [31–33]. Parametric models driven by algorithms for exploring kinetic facade design for daylighting performance [34], wind motion [35] and dynamic shading [36] have been reported.

1.2. Motivation

As described above, current models of responsive components lack the integration of variable geometry and do not include carrier component structures. To address these shortcomings, we propose the integration of geometric, structural and electronic subsystems supporting discrete adaptive motion within geometric constraints. Such a responsive component dynamically integrates environmental information in real time, and responds by changing its geometric state through structural integration, decomposable surface and open reprogrammable components. The design and assembly of a RCCE are based on the following features:

Support Geometry. The support grid and overall geometry should be defined by a carrier surface geometry [37]. The carrier surface is a self-supporting, integrated structural grid geometry, avoiding peripheral structural or cladding elements. The support system builds on our previous work on carrier component surfaces [38] and responsive carrier component envelopes [39].

Discretization. The carrier surface should be decomposable into an aggregation of discrete repeatable components. Self similar kinetic components are aggregated to minimize cost and complexity and provide a coherent integration between geometry/motion, responsive/electronic and physical/structural systems. The RCCE discretization builds on our previous work component aggregation including representative volume elements [38] and carrier component envelopes [28].

Open reprogrammable components. The carrier-component structure should support open ended reprogrammable sensor input–output for responsiveness to internal and external change parameters. Components should be reprogrammable and reconfigurable to allow a broader range of experiments with interactivity. The responsive methodology developed in [39,40] is used to develop the responsive/electronic subsystem of sensor networks.

In the next section, we present the design and assembly of a RCCE made feasible through advances in the integration of geometric, material and electronic subsystems as described above.

2. Materials and Methods

This section describes the geometry and control aspects of the the RCCE experimental prototype covering the three features identified above, namely support geometry, incorporation of adaptive motion through discrete components and reprogrammable interaction.

The project was undertaken in collaboration between IdeaFactory (YunTech) and CodeLab (Curtin University) under the supervision of Carl Yu (OneWork.io), a startup company with expertise on Internet of Things (IoT). The Curtin team focused on the design of the carrier surface and kinetic component prototyping while The YunTech team focused on the fabrication and making of the RCCE and put together prototypes, improved aspects of geometry and code and took part in the final assembly. All three teams cooperated remotely on design and fabrication of the prototype, improved aspects of kinetic geometry, material development code and took part in the final assembly. To accomplish the task, three teams worked simultaneously in different locations, namely structure/carrier component design, fabrication process refinement and digital fabrication factory culminating in a three-day public installation assembly at the Shenzhen MakerFaire Exhibition. In the next sections, we describe the design and fabrication of the RCCE.

2.1. Carrier Component Geometry

The design geometry of the RCCE was developed with both top-down and bottom-up approaches using modeling and scripting techniques. The structural grid fabrication was based on the approach presented by Andres Sevtsuk and Raul Kavlo in the SUTD pavilion [41]. The design is a modular, component-driven bottom-up system assembly exploring full-scale material and electronic subsystems. The top-down approach involved creation of a grid shell surface structure for carrier surface. The outcomes of the carrier component design process are summarized in Figure 1.

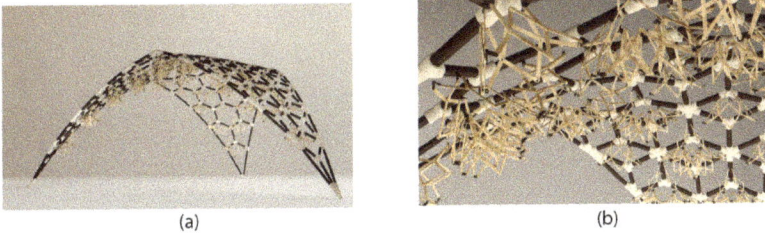

Figure 1. Carrier component structure: (**a**) detailed design of a discrete carrier surface as a support structure; and (**b**) aggregation of a single adaptive component over the surface.

This presents significant challenges in terms of assembly of complex interactive components. One of the challenges to the introduction of a curved structural grid was variation in the geometry of parts. In the bottom-up approach, the component scissor truss assembly was derived from four sides of a dodecahedron in Grasshopper. The movement of the arms of the scissors was simulated using a spring system in Kangaroo (Figure 2).

Figure 2. (**a**) Design rendition of responsive envelope on carrier surface; and (**b**) adaptive component triggering closed and open states through motion.

2.2. Adaptive Component

The components followed rules of local independence and global correspondence to form iterative responses and spatial configurations based on various local inputs. The moveable components edre based on the Hoberman Sphere and provided an adaptive assembly constrained to open and close in response to sensing change in the environment. The next step was to standardize kinetic component parts, namely the guiding rods to fit the curving surface. The self-organization was constrained by surface geometry on which the component was aggregated. Expansion and retraction of a symmetric polar array based on the Hobermann sphere was investigated. Parts had a degree of freedom in a complete system and worked independently while responding to their neighbors effecting changes on the larger scales, as simulated in Figure 3.

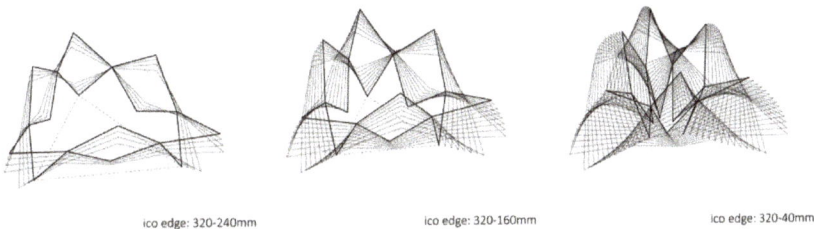

ico edge: 320-240mm   ico edge: 320-160mm   ico edge: 320-40mm

Figure 3. The design development of the adaptive component simulation of the scissor truss motion using a spring system simulation.

We tested the kinetic responsiveness of the RCCE with material constraints and simulated responses by connecting the adaptive components with programmable input and behavior.

This allowed for less deterministic part-to-whole relationships to emerge from interactions of local protocols and a multitude of external stimuli. In the current prototype, components in RCCE are comprised of scissor truss assembly based on Hoberman sphere guided by sliding arms and allow a single degree of freedom in a symmetrically polar array, as shown in Figure 4.

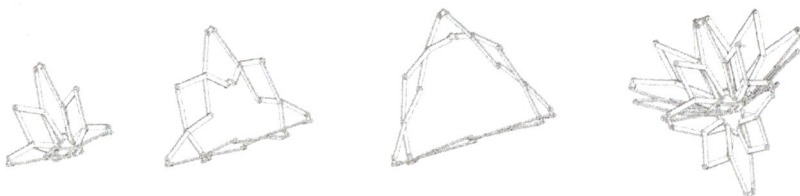

Figure 4. Detailed design modeling of polar array assemblies based on the expansion and contraction principles of a Hobermann sphere.

A standardization script was developed that fitted equilateral triangles using geometric transformation per face. Each component was comprised of a scissor truss assembly based on Hoberman sphere guided by sliding arms and allowed a single degree of freedom in a symmetrically polar array. The complexity of the assembly and its kinetic actuation, variation of parameters and scale were tested through material fabrication, as shown in Figure 5.

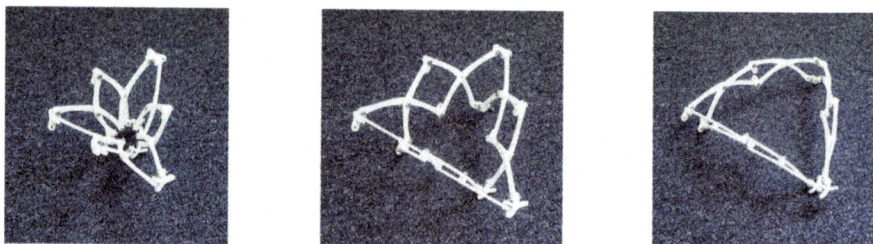

Figure 5. Physical prototype models with variational play to understand the stages of motion from closed to open.

The dynamic component was based on a planar surface tessellation with equilaterals and separated from the structural support grid. The adaptive component configuration prototype model was based on the scissor truss motion of a Hobermann sphere. This design compromise simplified the kinetic component and enabled the delivery of local interactions without structural ramifications. The self-organization was constrained by surface geometry, on which the component was aggregated. Parts had a degree of freedom in a complete system and worked independently while responding to the neighbors effecting changes on the larger scales. This allowed for a less deterministic part-to-whole relationship to emerge from interactions of local protocols and a multitude of external stimuli (Figure 6).

A few 3D-printed components were used to connect assembly at key intersections with unique angles. In this case, 3D printing was used as part of the combinatorial approach to construction where many different materials and parts came together as oppose to creating complex, continuous forms (Figure 7).

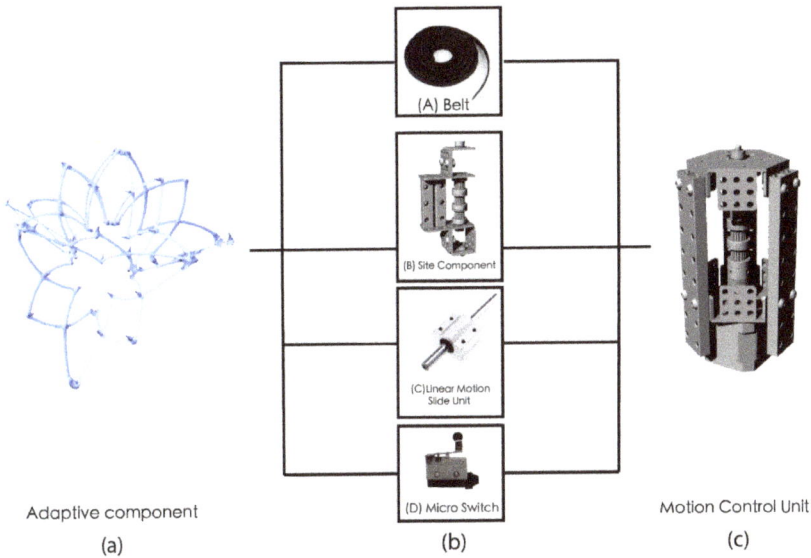

Figure 6. (**a**) Adaptive component configuration prototype model based on the scissor truss; (**b**) MakeBlock components for kinetic motion; and (**c**) central motion control unit.

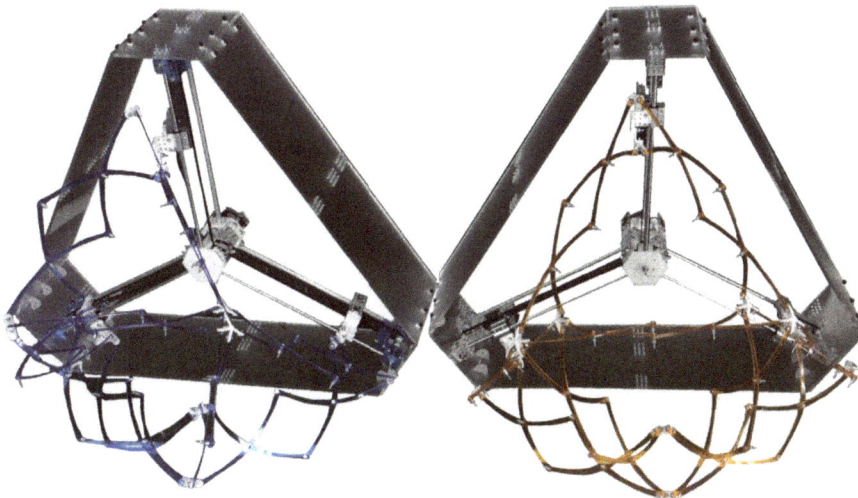

Figure 7. Early component configuration prototype models for kinetic components based on the scissor truss.

Early prototypes contained springs that kept component in tension while the final prototype used timing belts and motors to control position more precisely and eliminated kinetic energy, which made motion more unstable given relatively large number of parts per truss. The design used a number of standard industrial components such as steel shafts and connector brackets to reduce construction time and cost (Figure 7).

2.3. Kinetic Component Fabrication

The basic unit was built on a triangle containing tetrahedral scissor truss geometry (Figure 8). Structurally, the components were made of three main parts: (1) the structural frame that housed all the parts and connects it to its neighbors; (2) a guiding rail assembly that contained all sensing, actuating and logic components; and (3) kinetic scissor truss assembly. The 34:1 motor connected to three linear belts that guided the base points of the truss. There was a binary switch that communicated to the script terminal position of the truss under maximum contraction. A flexible multi-axis 3D-printed connector allowed us to connect the kinetic scissor truss to a planar rail frame that could be animated by linear motion.

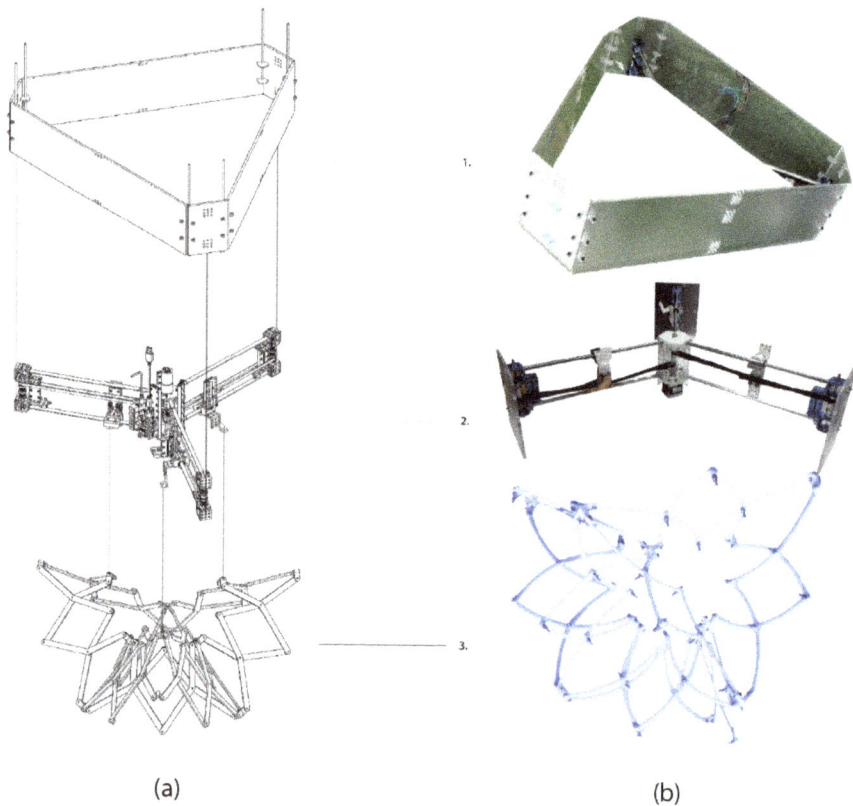

(a)   (b)

Figure 8. (a) RCCE design was done in three subcomponents comprising connectors, motion bars and scissor truss configuration; and (b) view of the final adaptive component assembly and fabrication.

An ultrasonic sensor (HC-SR04) driven by Arduino microcontroller detected proximity of people in front of it with a maximum range of about 5 m. Upon detection, it activates a geared motor connected to the kinetic truss through linear motion and began contracting it. Once the sensor stopped detecting or the collision switch on one of the arms was hit with the truss arm, the script pauses and proceeded to return the mechanism to its extended configuration.The control system used standard Arduino IDE written in Cal frame. The microcontroller was connected to the motor, ultrasonic sensor, an LED strip and a switch to determine position of the kinetic truss. This allowed components to "sense" presence and proximity and respond by retraction or contraction. The precise timing allowed for real-time control and synchronization of motor positions in relation to external stimuli.

2.4. Prototyping and Assembly

The prototyping process used MakeBlock [42] components for standard mechanical parts and electronic modules such as sensors, control boards, motor drivers and communicators. The fabrication process was undertaken experimentally using a trial-and-error process, with the fabrication team iteratively testing all control mechanism and parts. The laser cut process took the 3D design to create 2D sheet layouts via Grasshopper scripts. The non-standard parts were laser-cut Plexiglas elements for peripheral structural arms and 3D-printed components for kinetic mechanism and connectors not available from the MakeBlock component library. The 3D printer for the kinetic mechanism components used two materials, PLA and wax, for different parts due to different requirements for assembly. For instance, the gear for timely belt control required more accurate model, thus we used wax 3D print, but it is softer. The central part of gear for motor to spin needed a tighter connection for headless setscrew and metal shaft pin, thus we decided to use PLA.

A few 3D-printed components were used to connect assembly at key intersections with unique angles. In this case, 3D printing was used as part of the combinatorial approach to construction where many different materials and parts came together as opposed to creating complex, continuous forms (Figure 9).

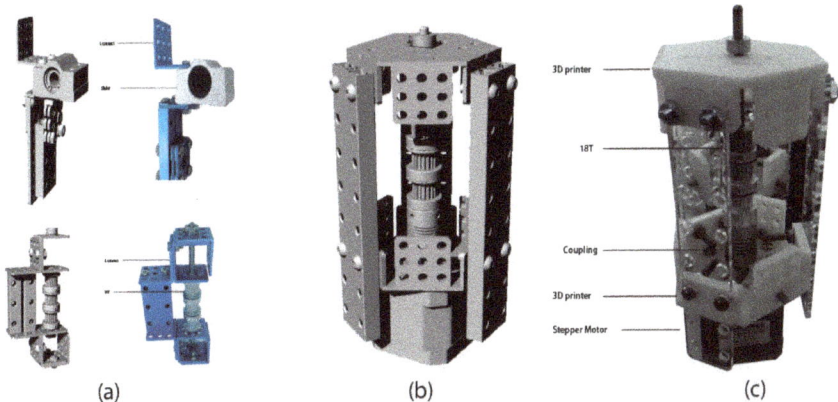

Figure 9. (**a**) Three subcomponents comprising connectors, motion bars and scissor truss configuration; (**b**) RCCE motion control unit; and (**c**) component assembly and fabrication.

Over the period of three weeks, the teams communicated and exchanged 3D models, scripts and other documentation relating to prototype construction. The Curtin team assembled design documentation, while Yun Tech team prototyped various design iterations using digital fabrication equipment at their facilities.The fabrication process took places in IDF at Yunlin, Taiwan and test assembly process in factory first and then all 16 sets were assembled on site in Shenzhen. Many factors affected the prototyping and assembly process:

- Tolerance and accuracy of parts via fabrication machine
- Mechanical control of movement via electronic boards
- Mechanical movement and vibration
- Parts assembly direction, thickness of material and its durability
- Friction between different parts

These factors affected the assembly and fabrication process and required the testing of multiple fabrication and assembly methods iteratively in a very short period time. The design team also needed to change the original design to accommodate the difficulties of fabrication (Figure 10). An agile design process methodology provided transparent workflow to the entire team to understand the status of different activities as the project progressed. It also held team members accountable for what they were assigned and documented feedback directly from all remote teams.

Figure 10. RCCE component prototype variations.

2.5. Exhibition

Finally, a concrete situation from practice was presented, where 16 fully functional components of the adaptive component were assembled and tested as part of an interactive public placemaking installation at the Shenzhen MakerFaire Exhibition. The aim of the RCCE was to develop an interactive media between people and space, as described in [1]. The RCCE in this sense represented an interactive media that received/sensed the interactive behaviors of people and reacted back into space. The behavior of people and the performative aspects of responsive installations were collected as video recordings on site and are provided in the Supplementary Materials to reflect the interaction between people and space. Design and research were conducted on multiple aspects of the process, remotely communicating between the teams in Shenzhen, Taiwan and Perth, Australia. The three teams worked simultaneously in different locations to develop structure/carrier component design, fabrication process refinement and digital fabrication, respectively. The full scale RCCE prototype was assembled in Shenzhen culminating in a three-day public installation assembly at the Shenzhen MakerFaire Exhibition.

Each component of the RCCE installation was an autonomous mechanism with an ultrasonic sensor (HC-SR04) driven by Arduino microcontroller. The sensor detected the proximity of objects (in this case people) in front of it with a maximum range of about 5 m. Upon detection, it activated a geared motor connected to the kinetic truss through linear motion and began contracting it. Once the sensor stopped detecting or the collision switch on one of the arms was hit with the truss arm, the script paused and proceeded to return the mechanism to its extended configuration (Figure 11).

Figure 11. Full-scale RCCE prototype: Installation of RCCE on site at Shenzhen MakerFaire Exhibition.

3. Results

The experimental prototype installation provided new results into the design and construction of an adaptive RCCE assembly comprising repeatable components that operate on computation protocols and exhibit responsive kinesis. The research addressed key issues underlying RCCE prototypes using a design and fabrication process integrated with technical subsystems. The findings of the study cover system design and planning, choice of fabrication and assembly methods and incorporation of dynamic forms. First, the outcomes of our research propose that carrier component structures are a useful abstraction for the development of responsive architecture. They permit the design and aggregation of repeatable components as well as provide a metaphor for allowing discrete components to respond either as a single element or in combination. Second, detailed design, fabrication and assembly of a single adaptive component is a critical level of abstraction for addressing issues of structure, material complexity, analysis of motion and electronic control and sensing.

Finally, the public installation of an RCCE prototype at the Shenzhen MakerFaire Exhibition demonstrated the potential role of responsive architecture in the public domain. During the three-day exhibition, the installation performed without failure and thousands of citizens interacted with the kinetic motion. These activities were recorded using video and samples are provided in the Supplementary Materials. These outcomes are summarized in Figure 12. The results of the experiment for the development of a responsive skin prototype are as follows.

Multi-scalar structure. A multi-scalar approach is needed for the development of the carrier surface structure and its subsequent articulation into discrete components requiring geometric resolution at two scales. The tessellation of the carrier surface must be rationalized to account for both part-to-whole relationships as well as control of component variation and scaling. In the RCCE, we separated the structural grid from the component aggregation to achieve this.

Digital to Physical translation. The selection of materials and their tectonic properties for components and connections is critical for responsive architecture. The digital-to-physical translation from rationalized geometry to material requires a consideration of their assembly.

Sensor Design. Sensor density, an important element in interaction design, is the resolution–density trade-off in the type of sensor components. Electronic components and sensors require careful consideration of the interaction logic, input–output behaviors and sizing and resolution of components. Denser sensor arrays produce greater resolution but create more complexity in terms of set up and fabrication as well as driving up costs.

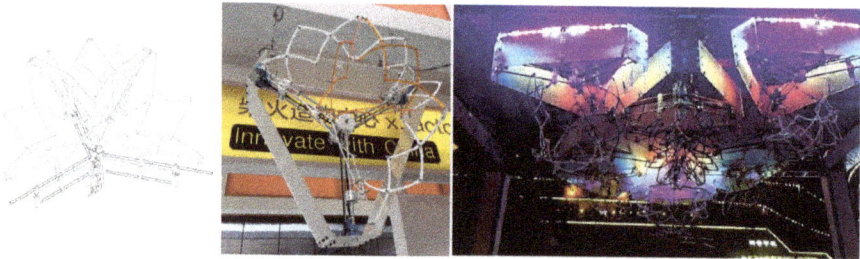

Figure 12. The outcomes of our research: (**left**) design research into the carrier component structures; (**middle**) detailed design, fabrication and assembly of a single adaptive component; and (**right**) public installation of an RCCE prototype at the Shenzhen MakerFaire Exhibition.

The protocols developed in the RCCE highlight the opportunities and consequences of how local adaptive components relate to the whole carrier envelope with multiple constraints and scale considerations.

4. Discussion

Responsive components are not a new phenomenon in architecture. Responsive architecture can be understood as any building or building component designed for adaptation to change. The RCCE is an experiment in responsive architectural design that gives us insight into construction of complex architectural assemblies incorporating structural, material, kinetic and electronic subsystems. RCCE is a work in progress and, due to its complexity and wide range of interdisciplinary knowledge required, will require expanded collaboration in the future.

The main problems encountered are structural and control issues, and the need to script and iteratively test a largely bottom-up system. Although the behavior is emergent, the form remains static. This problem will be addressed in further design iterations to control the local parameters influencing the surface tessellation and have more control over aggregation of local geometry into a larger whole. This could be expanded upon by looking at the way components connect, degrees of freedom allowed per component and the way cladding layer can be arranged more continuously rather than as a collection of disparate kinetic scissor truss assemblies. The future implementation will include communication protocol between components to allow for coordinated transformation of the structure. This tradition of design suggests that new ways of thinking about responsive components have their roots in a series of precedents that respond to change, such as the light sensitive apertures in the Institut du Monde Arabe in Paris, and they have the capacity to transform the way that architecture is experienced.

In conclusion, the design and assembly of adaptive components present new insights into designing sensitive, creative, adaptable, and sustainable architecture. Negroponte proposed that advances in artificial intelligence and the miniaturization of components would result in buildings capable of intelligently recognizing the activities of their users and responding to their needs, as well as changes in the external and internal environment [9]. There is a continuity in the design of automated building components, such as presented in this paper, and the long tradition of design in manually operated responsive mechanisms and static architectural elements that modulate inside-outside conditions. Meagher's study of the Maison de Verre shows how mechanical adjustment of components can achieve such a poetic responsiveness [2].

Supplementary Materials: The following are available online at http://www.mdpi.com/2075-5309/9/4/84/s1.

Author Contributions: Conceptualization, T.-W.C. and S.D.; methodology, T.-W.C.; software development, S.D.; validation, T.-W.C. and H.-Y.H.; formal analysis, all authors; writing—original draft preparation, S.D.; writing—review and editing, T.-W.C.; visualization, H.-Y.H.; supervision, T.-W.C.; project administration, H.-Y.H.; and funding acquisition, T.-W.C.

Funding: This research was funded by Idea Factory, YunTech, Taiwan and Curtin University, Australia.

Acknowledgments: The project was undertaken in collaboration between Idea Factory (Yunlin, Taiwan) and Code Lab (Perth, Australia) under the supervision of OneWork.io (Taipei, Taiwan), a startup company with expertise on Internet of Things (IoT).The authors acknowledge the contributions of the team leader of the project "Dynamic Cloud" Huei-Sheng Yu (onework.io) and the following participants: Andrei Smolik and Rex Auyeung (Curtin University); Tsai-Ling Hsieh, Yeong-Shenn Lee, Yun-Ru Chen and Chun-Yen Chen (Idea Factory, YunTech); Wayne Lin and Monica Shen (x.factory).

Conflicts of Interest: The authors declare no conflict of interest.

References

1. Bloomer, K.C.; Moore, C.W.; Yudell, R.J.; Yudell, B. *Body, Memory, and Architecture*; Yale University Press: New Haven, CT, USA, 1977.

2. Meagher, M. Designing for change: The poetic potential of responsive architecture. *Front. Arch. Res.* **2015**, *4*, 159–165. [CrossRef]

3. Ito, T. *Blurring Architecture, 1971–2005: Rethinking the Relationship between Architecture and the Media*; Art Books International London Milan: Charta, Colombia, 1999.

4. Goulthorpe, M.; Burry, M.; Dunlop, G. Aegis Hyposurface©: The Bordering of University and Practice. In Proceedings of the Twenty First Annual Conference of the Association for Computer-Aided Design in Architecture, ACADIA, CUMINCAD, New York, NY, USA, 11–14 October 2001; pp. 334–349.

5. Mignonneau, L.; Sommerer, C. Media Facades as Architectural Interfaces. In *The Art and Science of Interface and Interaction Design*; Sommerer, C., Jain, L.C., Mignonneau, L., Eds.; Studies in Computational Intelligence; Springer: Berlin/Heidelberg, Germany, 2008; pp. 93–104, doi:10.1007/978-3-540-79870-5_6.

6. Schittich, C. *Building Skins: Concepts, Layers, Materials*; Edition Detail; Birkhäuser: Munich, Germany, 2001.

7. Wigginton, M.; Harris, J. *Intelligent Skins*; Routledge: Abingdon, UK, 2013.

8. Tomitsch, M.; Grechenig, T.; Moere, A.V.; Renan, S. Information Sky: Exploring the Visualization of Information on Architectural Ceilings. In Proceedings of the 2008 12th International Conference Information Visualisation, London, UK, 9–11 July 2008; pp. 100–105, doi:10.1109/IV.2008.81. [CrossRef]

9. Sterk, T.d. Building upon Negroponte: A hybridized model of control suitable for responsive architecture. *Autom. Constr.* **2005**, *14*, 225–232, doi:10.1016/j.autcon.2004.07.003. [CrossRef]

10. Desai, A. Designing Building Skins. Master's Thesis, Dept. of Architecture, Massachusetts Institute of Technology, Cambridge, MA, USA, 1992.

11. Srisuwan, T. Fabric Façade: An Intelligent Skin. *Int. J. Build. Urban Inter. Landsc. Technol. BUILT* **2017**, *9*, 7–13.

12. Koudlai, A. *In and Out of Control: An Exercise in the Design of LOW-tech/High Effect Responsive Building Facades*; State University of New York at Buffalo: Buffalo, NY, USA, 2016.

13. Capeluto, G.; Ochoa, C.E. *Intelligent Envelopes for High-Performance Buildings: Design and Strategy*; Springer: Berlin, Germany, 2016.

14. Pottmann, H. *Architectural Geometry*; Bentley Institute Press: Exton, PA, USA, 2007.

15. Hudson, R. Parametric Development of Problem Descriptions. *Int. J. Arch. Comput.* **2009**, *7*, 199–216. [CrossRef]

16. Woodbury, R. *Elements of Parametric Design*; Routledge: London, UK; New York, NY, USA, 2010.

17. Woodbury, R.; Burrow, A.; Datta, S.; Chang, T.W. Typed feature structures and design space exploration. *Artif. Intell. Eng. Des. Anal. Manuf.* **1999**, *13*, 287–302. [CrossRef]

18. Aish, R.; Woodbury, R. Multi-level Interaction in Parametric Design. In *Smart Graphics*; Butz, A., Fisher, B., Krüger, A., Olivier, P., Eds.; Number 3638 in Lecture Notes in Computer Science; Springer: Berlin/Heidelberg, Germany, 2005; pp. 151–162.
19. Jeng, T., Designing a ubiquitous smart space of the future: The principle of mapping. In *Design Computing and Cognition'04*; Springer: Berlin, Germany, 2004; pp. 579–592.
20. Kim, D.Y.; Kim, S.A. An exploratory model on the usability of a prototyping-process for designing of Smart Building Envelopes. *Autom. Constr.* **2017**, *81*, 389–400. [CrossRef]
21. Datta, S. Modeling dialogue with mixed initiative in design space exploration. *Artif. Intell. Eng. Des. Anal. Manuf.* **2006**, *20*, 129–142, doi:10.1017/S0890060406060124. [CrossRef]
22. Jiang, H.; Chang, T.W.; Liu, C.L. Musical Skin: A Dynamic Interface for Musical Performance. In *Human-Computer Interaction. Interaction Techniques and Environments*; Jacko, J., Ed.; Lecture Notes in Computer Science; Springer: Berlin/Heidelberg, Germany, 2011; Volume 6762, pp. 53–61, doi:10.1007/978-3-642-21605-3_6.
23. Chang, T.W.; Datta, S.; Lai, I.C. Modelling Distributed Interaction with Dynamic Agent Role Interplay System. *Int. J. Digit. Media Des.* **2016**, *8*, 1–14.
24. Seely, J.C. Digital Fabrication in the Architectural Design Process. Ph.D. Thesis, MIT, Cambridge, MA, USA, 2004.
25. Kolarevic, B. Back to the Future: Performative Architecture. *Int. J. Arch. Comput.* **2004**, *2*, 43–50. [CrossRef]
26. Huang, H.Y.; Chang, T.W.; Wu, Y.S.; Chen, J.Y. Collective Fabrication A Responsive Dynamic Skin Design Case. In Proceedings of the 22nd International Conference on Computer Aided Architectural Design Research in Asia; Janssen, P., Loh, P., Raonic, A., Schnabel, M.A., Eds.; CAADRIA: Singapore, 2017; pp. 99–100.
27. Foged, I.W.; Kirkegaard, P.H.; Christensen, J.T.; Jensen, M.B.; Poulsen, E.S. Shape Control of Responsive Building Envelopes. In Proceedings of the International Symposium of the International Association for Shell and Spatial Structures (IASS): Spatial Structures—Temporary and permanent, Shanghai, China, 8–12 November 2010; IASS, China Architecture and Building Press: Beijing, China, 2010; pp. 2602–2609.
28. Datta, S.; Hanafin, S.; Woodbury, R.F. Responsive envelope tessellation and stochastic rotation of 4-fold penttiles. *Front. Arch. Res.* **2014**, *3*, 192–198, doi:10.1016/j.foar.2014.03.002. [CrossRef]
29. Pesenti, M.; Masera, G.; Fiorito, F.; Sauchelli, M. Kinetic solar skin: A responsive folding technique. *Energy Procedia* **2015**, *70*, 661–672
30. Chang, T.W.; Jiang, H.; Chen, S.H.; Datta, S. Dynamic Skin: Interacting with Space. In Proceedings of the 17th International Conference on Computer Aided Architectural Design Research in Asia, CAADRIA, Chennai, India, 25–28 April 2012; pp. 89–98.
31. Chen, X. Interactive Pavillions: Responsive Transformation of Structure Systems. Thesis Prep, School of Architecture, Syracuse University, Syracuse, NY, USA, 2015.
32. Shaviv, E. Integrating energy consciousness in the design process. *Autom. Constr.* **1999**, *8*, 463–472, doi:10.1016/S0926-5805(98)00101-0. [CrossRef]
33. Radford, A.D.; Gero, J.S. *Design by Optimization in Architecture, Building, and Construction*; Van Nostrand Reinhold: New York, NY, USA, 1988.
34. El Sheikh, M.M. *Intelligent Building Skins: Parametric-Based Algorithm for Kinetic Facades Design and Daylighting Performance Integration*; University of Southern California: Los Angeles, CA, USA, 2011.
35. Datta, S.; Hanafin, S.; Pitts, G. Experiments with stochastic processes: Facade subdivision based on wind motion. *Int. J. Arch. Comput.* **2009**, *7*, 389–402. [CrossRef]
36. Chang, T.W.; Datta, S. Dynamic Envelopes: Structural and Interactive transformations. In Proceedings of the 7th CUTSE Conference, Sarawak, Malaysia, 1 August 2012; pp. 89–98.
37. Pitts, G.; Datta, S. Parametric Modelling of Architectural Surfaces. In Proceedings of the 14th International Conference on Computer Aided Architectural Design Research in Asia, Yunlin, Taiwan, 22–25 April 2009; Chang, T.W., Champion, E., Chien, S.F., Chiou, S.C., Eds.; CAADRIA: Yunlin, Taiwan, 2009; pp. 635–644.
38. Hanafin, S.; Pitts, G.; Datta, S. Non-Deterministic Exploration through Parametric Design. *Int. J. Arch. Comput.* **2009**, *7*, 605–622. [CrossRef]
39. Smolik, A.; Chang, T.W.; Datta, S. Prototyping Responsive Carrier-Component Envelopes. In Proceedings of the 22nd CAADRIA Conference, Suzhou, China, 5–8 April 2017; Janssen, P., Loh, P., Raonic, A., Schnabel, M.A., Eds.; CAADRIA: Suzhou, China, 2017; pp. 521–528.

40. Datta, S.; Andrei, S.; Chang, T.W. Responsive Interaction in Dynamic Envelopes with Mesh Tessellation. In *CAADence in Architecture International Workshop and Conference*; Szoboszlai, M., Ed.; Dept. of Architecture and Interior Architecture: Budapest, Hungary, 2016; pp. 241–246.

41. Sevtsuk, A.; Kalvo, R. A freeform surface fabrication method with 2D cutting. *Simul. Ser.* **2014**, *46*, 138–145.

42. Makeblock Co., Ltd. Makeblock. Available online: http://learn.makeblock.com/en/ (accessed on 12 October 2016).

MDPI

St. Alban-Anlage 66

4052 Basel

Switzerland

Tel. +41 61 683 77 34

Fax +41 61 302 89 18

www.mdpi.com

Buildings Editorial Office

E-mail: buildings@mdpi.com

www.mdpi.com/journal/buildings